Coherent Wireless Power Charging and Data Transfer for Electric Vehicles

Authored by

Chih-Cheng Huang
National Space Organization
Hsinchu City
Taiwan

Chun-Liang Lin
Department of Electrical Engineering
National Chung Hsing University
Taichung City, Taiwan

Coherent Wireless Power Charging and Data Transfer for Electric Vehicles

Authors: Chih-Cheng Huang & Chun-Liang Lin

ISBN (Online): 978-1-68108-946-1

ISBN (Print): 978-1-68108-947-8

ISBN (Paperback): 978-1-68108-948-5

Bentham Science Publishers Ltd.
Executive Suite Y - 2
PO Box 7917, Saif Zone
Sharjah, U.A.E.
Email: subscriptions@benthamscience.net

CONTENTS

PREFACE

Due to the global awareness regarding environmental protection, instead of high-emission, low-efficiency automobiles, low-emission, high-efficiency vehicles are being manufactured. An electric vehicle uses the energy stored in the battery to drive high-performance electric motors. High-efficiency electric motors have inherently low-speed and high-torque that make electric vehicles ideal for driving in an urban area; therefore, the electric vehicle and hybrid electric vehicle are favored in the household market. The electric vehicle needs charging before its battery is exhausted. There are two technologies to charge electric vehicle batteries, conductive (wired) and wireless charging. Whether conductive or wireless charging, the recent batteries take 0.5-12 hours charging time. The grid transfers conductive charging power to the battery through a wire. It has clear benefits, but it must meet the “wired charging four necessary conditions,” which will be discussed later. Wireless charging transfers power through the magnetic field between two aligned coil pads. It does not rely on those conditions. The only drawback the wireless charging has is that two conductive coil pads must align for efficient power transfer. The manufacturer has to provide a parking guidance device to assist the user in parking the vehicle and align two coil pads. This tutorial book introduces a wireless charging method that does not only have massive power, compact size, and high transferring efficiency but provides two-frequency bandwidth and bidirectional communication using a data-attached mechanism without resorting to other RF devices. The novel communication provides charging and billing information and monitors safety during the charging period.

The conductive charging is quite mature and compliant with regulations of SAE J2931/1, J1772, SAE J2836/6, IEEE 802.11, etc. Its advantages are high efficiency of about 88-91% on full load and being straightforward. However, the safety during the charging period must satisfy “wired charging four necessary conditions.” The connector must be tightly linked, there must be a stable power source, the vehicle must be grounded, and the vehicle must be stationary. . As time progresses, the aging effect of the components can occur. As a result, wire and connectors must be maintained; uncertain grounding may lead to leakage or sparking; vehicle movement poses a risk during charging. As depicted, the conductive charging will bring an extra cost and risk to the users, however, these disadvantages do not occur on the wireless charging approaches.

The wireless charging device includes two parts: the primary unit and the secondary unit. The former is on the grid side, which generates high-frequency current entering a coil pad, and the latter is on the vehicle side, which receives the power energy via a coil pad. The two pads are aligned together for high-efficiency power transfer. SAE J2954 is the industrial regulation, but it is currently in the drafting stage. Wireless power transfer is based on the principle of electromagnetic inductive or magnetic resonance. Inductive resonance has a wide range of applications such as an inductive heater in the kitchen with an efficiency of about 84%. The advantage of wireless charging by inductive technology is convenience; its efficiency may be up to 86%. Because there is no wiring, it does not need to consider “wired charging four necessary conditions.” The vehicle can even move or leave during the charging process.

This book shows a data attached technology to synchronize the power carry wave and magnetic field link between two inductive pads. Its advantages include the fact that data is not broadcasted but rather hidden within the carrier, and it is not susceptible to interference.This is quite suitable for wireless charging applications. The two frequency band includes regular data using the low-frequency band and emergency data using the high-frequency band. Information to be transferred may include vehicle ID, charging voltage, battery current and status, which let the primary unit know which car is being charged for billing and what is the charging status for safety. If an alarm goes off, such as for overvoltage or overcurrent, the emergency data will give feedback to the primary unit to stop charging immediately. The response time is faster than WiFi, Bluetooth, or ZigBee.

The reason for the long charging time is the chemical reaction process of redox in the battery. For example, an electric vehicle generally uses a lithium-ion battery as a power source. A cell is a basic electrochemical unit that contains the electrodes, separator, and electrolyte. Its charging process is that first, the lithium-ions escape from the positive electrode through the separator and the electrolyte. Finally, these lithium-ions diffuse to the negative electrode. The diffusion time becomes the key to charging speed. To enhance the charging rate, new electrode materials have been continuously discovered. For example, ultra-capacitor electrode materials like graphene, carbon nanotubes (CNT), activated carbon, and others rely on static electricity to charge and discharge the battery instead of chemical reactions. Therefore, the ultra-capacitor has fast charging and discharging speed and high power density, but its energy density is lower than the battery according to the same size. As depicted, the process of reducing the material size and increasing energy density will continue.

CONSENT FOR PUBLICATION

Not applicable.

CONFLICT OF INTEREST

The authors declare no conflict of interest, financial or otherwise.

ACKNOWLEDGEMENT

This book has been edited by Wallace Academic Editing.

Chih-Cheng Huang
National Space Organization
Hsinchu City
Taiwan

Chun-Liang Lin
Department of Electrical Engineering
National Chung Hsing University
Taichung City, Taiwan

INTRODUCTION

To focus on reducing emissions and improving fuel economy, automotive manufactories are developing Electric Vehicles (EV) to replace fuel and diesel vehicles starting from 2030~2040. The green power supply EVs to make a maximum environmental benefit, zero-emission, and lower pollution. However, changing climate occurs intermittently, which results in a spinning reserve of electric power. Fortunately, a stable electric storage system (ESS) may compensate for this problem.

Electromagnetic induction transmits power from the source to the load via air gap based on the Faraday theorem, Ampere theorem, and Maxwell equation. A power pad combines with a turned network (or compensation network), constituting a resonant circuit to transmit and receive more massive energy efficiently. As a result, it can efficiently transmit massive power with higher efficiency of up to 90%, the quality factor of 5~100, a coupling factor of 0.2~0.5, and a fundamental operating frequency range from hundreds of Hz to several 100 kHz through an acceptable air gap. These achievements are a credit to high-frequency semiconducting switching components with less power loss, and a tuned compensation network can compensate for power transmission loss due to lateral misalignment between coils. In addition, high-quality power pads can provide a higher magnetic flux to overcome the limitation of the air gap.

A static battery charging system with a heavy electric storage tank will definitely help the vehicle to achieve a longer traveling distance. However, the heavy electric storage system will encumber the vehicle and produce more pollution in the environment. On the contrary, a dynamic battery charging system can reduce the vehicle battery size and weight while increasing vehicle driving efficiency. However, the system needs to integrate various infrastructures such as the widespread battery charging stations.

Adaptive compensation of the magnetic inductive system (ACMIS), based on zero voltage switch (ZVS), transmitting electric power from a single-end inverter combined with a compensated network of parallel to parallel (P-P) type and an auto-tuning impedance of LC tank is introduced in this book. The issue of simultaneous power and data communication is covered. The coherent wireless data transferring scheme includes handshaking communication, a current trimming mechanism, and a data attached scheme that synchronizes with the power flow via magnetic link. The advantages are low cost and RF radiation and interference. In addition, it simultaneously carries feedback of the load side's message in real time. The experiment for a multifunctional contactless power flow of the G2V mode and bidirectional outer communication and inner communication with giant magnetoresistance (GMR) effect for car parking guidance is introduced. The experiment analyzes data transferring performance, including the current trimming method and data attached method, to evaluate data transmission quality according to the varying lateral offset, output power, and the air gap between two inductive power pads.

CHAPTER 1

Background

Abstract: Automotive manufactories are developing electric vehicles, such as hybrid electric, plug-in hybrid, battery-electric, and fuel cell electric vehicles in order to reduce emissions and improve fuel economy. Major advanced countries will ban fuel and diesel cars starting from 2030~2040. Electric, hybrid, and fuel cell vehicles have attracted more and more attention from automakers, governments, and customers. Research and development efforts have been focused on developing novel concepts, low-cost systems, and reliable hybrid electric powertrain.

This chapter reviews the present technologies of EVs in the range of Li-ion battery technology, drivetrain configuration, electric motor drives, power distribution system management with charging/discharging of EVs and conductive/wireless charger.

Keywords: Aggregator, Battery electric vehicle (BEV), Conductive/wireless charger, Drivetrain, Electric motor, Fuel cell electric vehicle (FCEV), Hybrid electric vehicle (HEV), Li-ion battery, Plug-in hybrid electric vehicle (PHEV), Power grid.

1.1. INTRODUCTION

The vehicles are propelled by Internal Combustion Engines (ICE), which emit carbon monoxide, carbon dioxide, nitrogen oxides, and hydrocarbons as a result of the fuel they burn [1, 2]. These results will lead to global climate change, such as rising sea levels, abnormal rainstorms, droughts, and increased desertification. These impacts will cause significant harm to the ecosystem, water and land resources, human socioeconomic activities, and life safety.

A total of 195 countries have adopted the first-ever universal and legally binding climate change agreement at the Paris Climate Conference in December 2015 [3]. This agreement issues a comprehensive action plan to put humankind on track to limit global warming below 2°C above the preindustrial levels. This ambitious plan requires a significant reduction in greenhouse gas emissions, starting from 2020. According to the International Energy Agency, the long-term concentration of greenhouse gases in the atmosphere must be limited to about 450 parts per million of carbon-dioxide equivalent [4].

Chih-Cheng Huang & Chun-Liang Lin

According to the Paris climate agreement, major advanced countries will ban fuel and diesel cars starting from 2030~2040, such as Germany (2030), China (2040), the United States (2030), the United Kingdom (2040), France (2040), India (2030), *etc.* This will force consumers to purchase vehicles with high efficiency and low pollution [5]. Compared with traditional cars, electric vehicles not only use energy more efficiently but also have a wide range of sources of electricity, such as hydropower, thermal power, nuclear power, renewable energy, *etc.* There is a growing interest in hybrid electric vehicle (HEV), plug-in hybrid electric vehicle (PHEV), Fuel cell electric vehicle (FCEV), and battery electric vehicle (BEV) technologies because of their reduced fuel usage and greenhouse emissions [6 – 8].

HEV use more than one source of energy. The design of the propulsion system can be adapted to different output power curves to achieve higher efficiency. For example, ICE has a high efficiency; an electric motor (EM) has a higher efficiency than ICE while accelerating or decelerating frequently. Both sources of power operate with a controller to save fuel and reduce emissions. In addition, it can use a battery for regenerative braking, recovering kinetic energy to save energy when the vehicle is decelerating. It is favored by many consumers since it has minimal mileage concern and supports high efficiency. In 2018, HEVs accounted for around 3% of global car sales (2.8 million units), compared to 1.5% for battery electric vehicles (about 1.4 million units). Under increasingly stringent conditions of global vehicle emissions, the IHS Markit [9] predicts that the amount of HEV and BEV sold is less compared with ICE, as shown in Fig. (**1.1**).

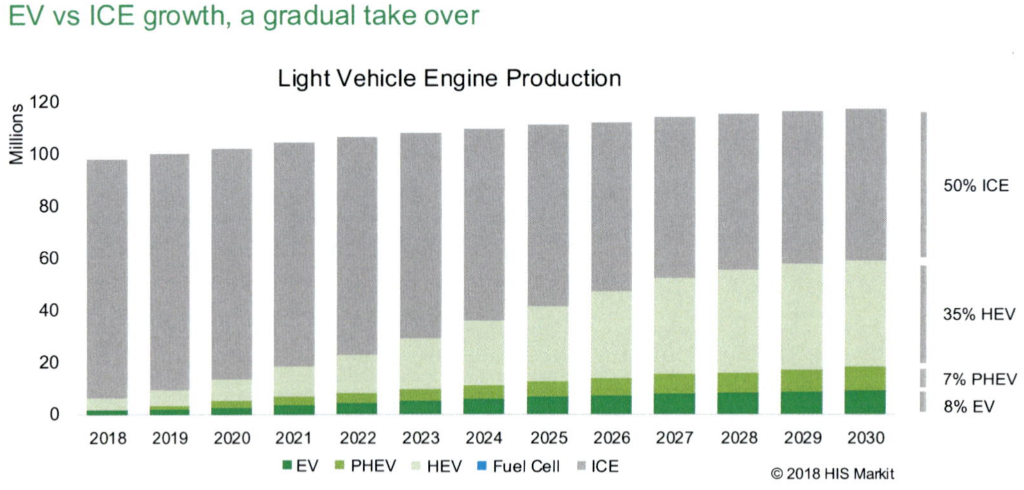

Fig. (1.1). The market prediction of EV, PHEV, HEV, Fuel cell and ICE.

Famous hybrid vehicles such as Toyota's Prius (PHEV) and related series have a battery capacity of about 1 to 9 kWh [10]. Although the mileage in the purely electric mode is very short, the overall fuel consumption can reach at least 20 km/l when the hybrid drivetrain is in parallel mode. In global sales, as of Oct. 2014, the sales volume is more than 7 million vehicles, which indicates that it is a fairly mature product.

1.2. Powertrain Architecture Classification

Energy consumption is a major assessment when comparing electric, hybrid, ICE, and FCVs vehicles. Several authors have provided well-to-wheels analyses that assess the global energy consumption of these types of vehicles [1, 8, 11]. HEVs and BEVs have been confirmed to consume less fuel and gas emission compared with ICE vehicles, even when electrical power is generated using petroleum. About FCEVs, the results are not so obvious in terms of the production of hydrogen.

HEVs are propelled by an ICE and an electric motor (EM) in series or parallel configurations. The ICE provides the vehicle an extended driving range, while the EM increases efficiency and fuel economy by regenerating energy during braking and storing excess energy from the ICE during coasting. ICE and EM are integrated as powertrains to achieve high performance by an intelligent control system. The default program in the intelligent control system considers ICE as suitable for steady-state operation and the electric motor suitable for frequently accelerating and decelerating operation. Fig. (**1.2**) shows a running vehicle whose load power can be divided into two elements [11]. One is average (steady) power, which is constant power, and another is dynamic power, which is zero average power. The intelligent control system assigns ICE and EM output power to wheels based on speed changes, power, and demand. One powertrain which favors steady-state operation, such as an ICE, can be used to supply the average power. On the other hand, another powertrain, such as an electric motor, can be used to supply dynamic power. The total energy output from the dynamic powertrain will be zero in a driving cycle. The dynamic power is provided by the electric traction system (traction motor and chemical batteries) to meet the peak power demand and recover the braking power.

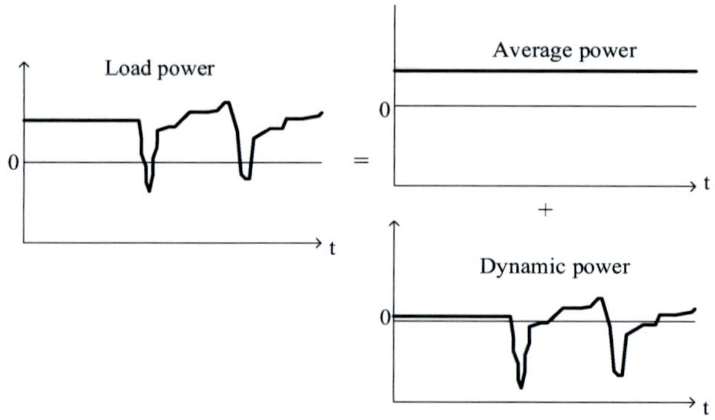

Fig. (1.2). Load power is decomposed into steady element and dynamic element.

The concept of hybrid powertrain can be implemented by different configurations (architectures) [11, 12], which are mentioned as follows:

1.2.1 Series Hybrid Drivetrain

The power progression is linear, resulting in a series hybrid-electric drivetrain. It runs on an internal combustion engine that drives a generator that only generates enough electricity to maintain steady-state operation (average power). Ideally, series HEV is an ICE-assisted EV that aims to extend the driving range. Depending on the battery capacity, if the battery is able to propel the vehicle, then it is classified as a full hybrid. If not, then it is called a mild hybrid. The converted electricity by ICE either charges the batteries or can bypass the battery to propel the wheels *via* the same electric motor and mechanical transmission. During braking, coasting, and transferring excess energy from the wheels to the batteries, energy is regenerated. The series hybrid drivetrain is illustrated in Fig. (**1.3**).

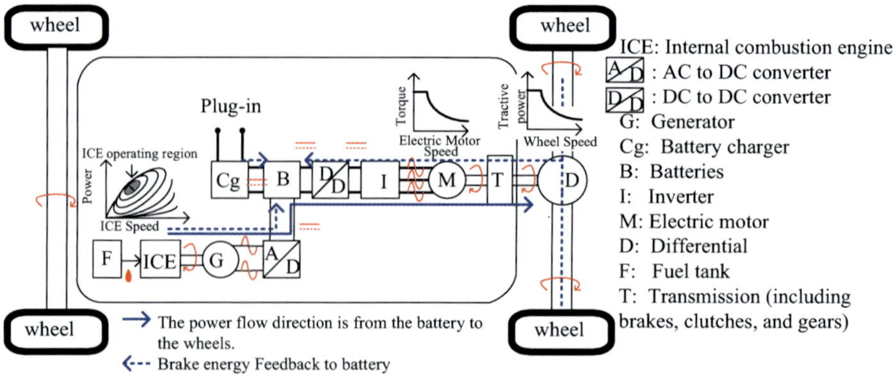

Fig. (1.3). Block diagram of series HEV.

Some of its advantages are: 1) Mechanical decoupling between the ICE and the driven wheels allows the ICE to operate in the narrow optimal region; 2) The engine generated by steady-state operating can increase efficiency and reduce emission; 3) nearly ideal torque-speed characteristic of electric motor [1, 13] makes complex multi-gear transmission unnecessary, which brings benefits such as increased cabin capacity, simplified mechanical maintenance, and no discontinuous power output during shift gear. Therefore, as a pure electric vehicle, the series hybrid is quiet and comfortable (BEV); 4) simple structure and drivetrain control.

Some of its disadvantages are: 1) series hybrid undergo multiple conversions, such as generator loss, battery charge loss, battery discharge loss, and motor loss, *etc.* Owing to the loss, the series hybrid is suitable for driving in urban areas, and it is useless on the highway; 2) the cascade structure leads to relatively low-efficiency ratings because all three motors are required. 3) A big traction motor is required since it is the only torque source of driven wheels.

Taking advantage of its simple structure, control, and low-speed and high-torque, this drivetrain is usually employed in heavy vehicles, such as passenger buses, military vehicles, and locomotives. In terms of fuel consumption, this system is suitable for frequently accelerating and decelerating operations.

1.2.2. Parallel Hybrid Drivetrain

Parallel Hybrid Drivetrain uses mechanical coupling in parallel ICE and electrical motor. A typical configuration of a parallel hybrid drivetrain system is illustrated in Fig. (**1.4**). Both the engine and the electric motor directly supply torques to the driven wheels that can be used separately or together to propel a vehicle. Conceptually, it is inherently an electric-assisted ICE vehicle for achieving lower emissions and fuel consumption. This drivetrain is classified as a mild hybrid. In the low-speed region, it employs a start-stop system [13] in the low-speed to reduce emissions and idling to increase fuel economy by using only an electric motor to drive the vehicle. In a steady speed region, only the internal combustion engine supplies high efficiency to drive vehicles. Both powertrains operate vehicles at the same time to provide comfort and pleasure when ascending or in a hurry. ICE power can be split to drive the vehicle and charge the batteries. When charging the batteries, the electric motor can be used as a generator to charge the battery through regenerative braking, coasting, or by absorbing power from the ICE when its output is greater than the power required for driving wheels.

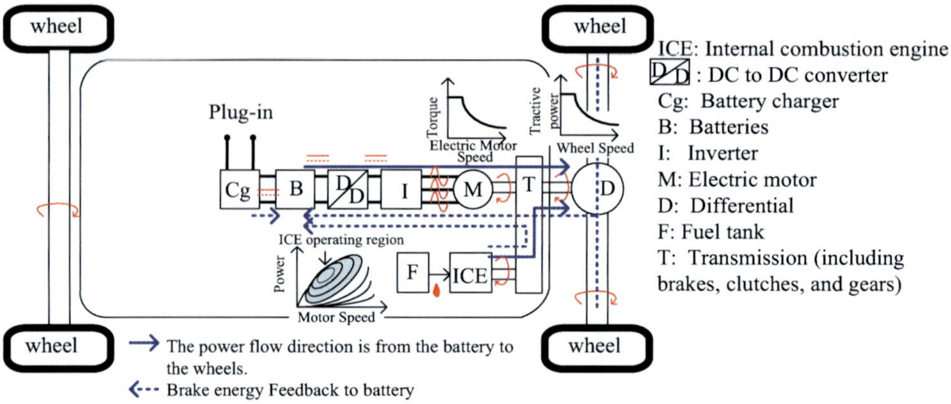

Fig. (1.4). Block diagram of parallel HEV.

Some of its advantages are: 1) As its powertrains are not cascaded, thus energy loss is less; 2) The compact structure of the drivetrain requires only smaller traction electric motor and ICE that can be used to obtain the same dynamic performance comparable with the series hybrid drivetrain; 3) lower cost and simple architecture.

Some of its disadvantages are: 1) As parallel hybrid drivetrains are the mechanical couplings between the engine and the driven wheels, thus the engine operation points cannot be fixed in a narrow speed region to obtain an optimal benefit; 2) clutches are often necessary for the start-stop system; 3) parallel hybrid vehicle needs a complex control strategy to drive wheels between ICE and electric motor.

The Honda Insight, Honda Civic, Ford Escape, and M. Benz S400, *etc.*, are some examples of parallel hybrid systems [13].

1.2.3. Series-parallel Hybrid Drivetrain Objectives

A typical configuration of the series-parallel drivetrain is shown in Fig. (**1.5**). In the figure, there is a planetary gear set in transmission, which decouples the engine speed from the wheel speed [13 – 15]. The motor and transmission shaft (TS) are connected to the planetary ring gear set (R), and the motor speed is proportional to the transmission shaft (vehicle speed). The ICE is connected to the carrier (C), and G/M is connected to the sum gear (S).

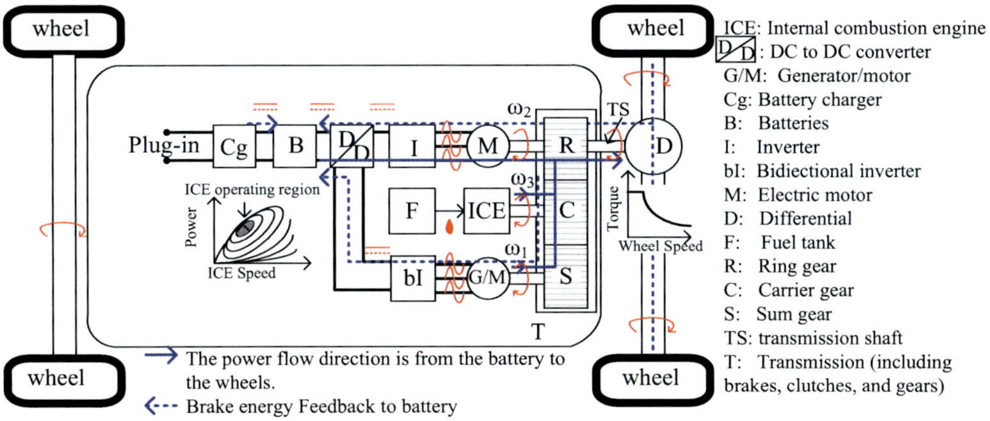

Fig. (1.5). Block diagram of serial-parallel HEV.

The planetary gear set is shown in Fig. (**1.6**). Thus the planetary gear unit, engine, and G/M constitute the series power flow route. When the G/M speed is negative (opposite direction *versus* the torque), the G/M operates in generating power. In this way, the ICE power is split into two paths; one path is transferred to the drivetrain, and the other generator charge batteries. When the G/M speed is positive, the G/M becomes a motor, and aids ICE to propel wheels.

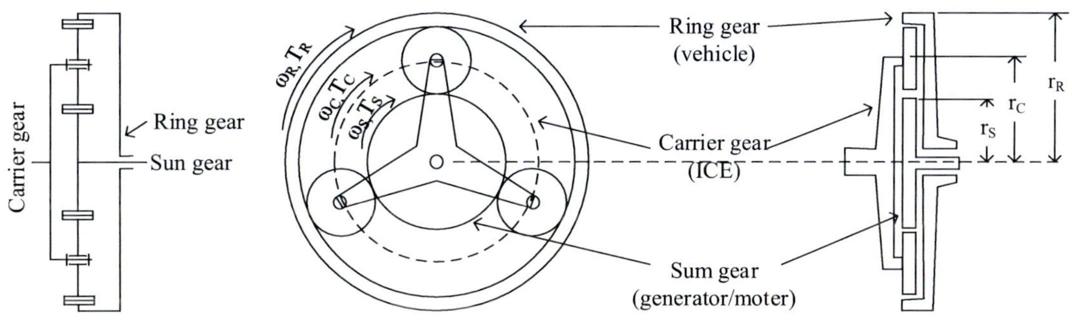

Fig. (1.6). Block diagram of the planetary gear set.

The speed and torque of an ICE are expressed as Eq. 1.1 and Eq. 1.2, respectively [11].

$$\omega_C = \frac{r_S}{2r_C}\omega_S + \frac{r_R}{2r_C}\omega_R \quad (1)$$

$$T_C = \frac{2r_C}{r_S}T_S = \frac{2r_C}{r_R}T_R \quad (2)$$

In Eq. (1.1), the ICE speed (ωC) is a weighted average of the speeds of the vehicle (ωR) and G/M (ωS). For any given vehicle speed, the G/M speed can be adjusted to choose the ICE speed under an optimal region by controlling the G/M (motor) speed.

The series-parallel configuration combines the advantages of series and parallel drivetrains. However, they still require three motors and a planetary gear set, which makes the powertrain somewhat complicated and costly. The well-known Toyota Prius employed this configuration. Another possible alternative to the planetary gear unit uses a combination of two concentric machines motor and G/M as a power-split device [16 - 18]. It reduces the architecture size and weight; the two machine sets can be merged, creating a single machine with a double rotor [19].

1.3. Electric Vehicles

1.3.1. BEV

BEV is the only powertrain that runs entirely on electricity, as illustrated in Fig. (**1.7**). It is exclusively powered by batteries or other electrical energy sources, and it is possible to achieve zero-emissions. However, the short driving range, as well as the high initial cost, large and bulky batteries, and long refueling time, has limited its use. The cost of electric vehicles reduced continuously, along with an increase in the market share. Investing in new batteries materials research provides high charging current, short charging time (80% full charging/ half hour); also, the cost of the batteries has fallen to less than $120/kWh, high power density (1–1.5kW/kg) [20], energy consumption between 117 and 268 Wh/km [21] and energy density (55-150Wh/kg) [20–21]. This energy density is quite low when compared to gasoline, which has an energy density of roughly 12000Wh/kg. Electric vehicles will bring more benefits in the future.

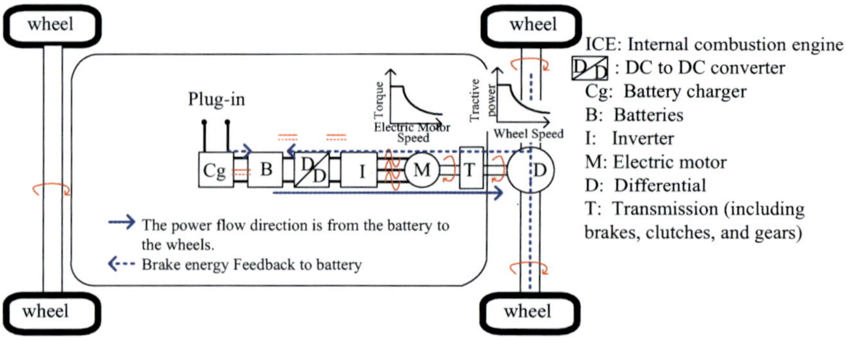

Fig. (1.7). Block diagram of BEV.

1.3.2. FCEV

FCEV uses fuel cells to generate electrical energy from hydrogen [22 - 24]. The electrical energy is either used to drive the vehicle or is stored in an energy storage device, such as a battery pack or ultra-capacitors. Because an FCEV can be equipped with batteries or super-capacitors, it can be regarded a form of BEV, as shown in Fig. (**1.8**). Since fuel cells generate electricity from a chemical reaction (isothermal) [25], they do not burn fuel, therefore, they do not produce pollutants. The byproduct of a hydrogen fuel cell is water. An FCEV provides quiet operation and more comfort for the ride. However, the production of the hydrogen fuel cell may result in emissions.

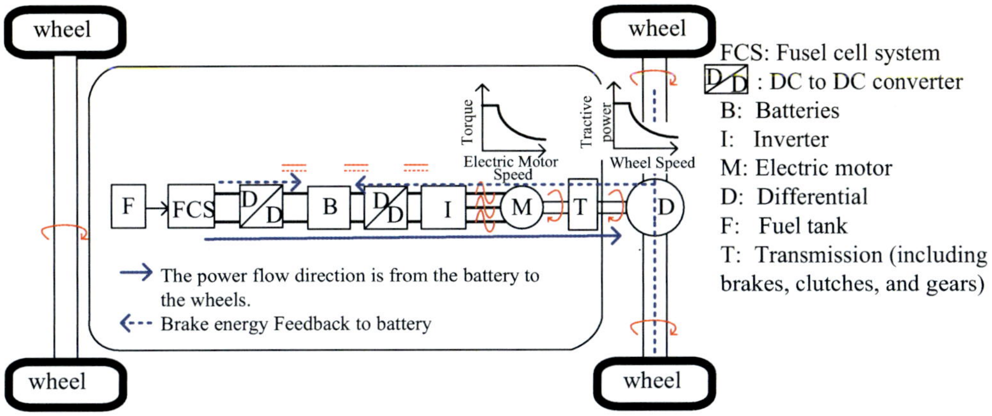

Fig. (1.8). Block diagram of FCEV.

Table **1.1** shows a comparison of the major characteristics of BEV, HEV, and FCEV.

Table 1.1. List of Major Characteristics of BEV, HEV, and FCEV.

Types of EVs	BEV	HEV	FCEV
Propulsion	√ Electric motor drives	√ Electric motor drives √ ICE	√ Electric motor drives
Energy storage	√ Battery √ Super-capacitor	√ Battery √ Super-capacitor √ Fossil or alternative fuels	√ Hydrogen tank √ Battery, super-capacitor needed to enhance power density
Energy system & infrastructure	√ Electric grid charging facilities	√ Gasoline stations √ Electric grid charging facilities for PHEV	√ Hydrogen √ Hydrogen production and transportation infrastructure

(Table 1.1) cont.....

Types of EVs	BEV	HEV	FCEV
Battery size [kWh]	35-80	1-10	1-10
Characteristics	√ Zero-emission √ Range anxiety √ Lower travel cost √ High initial cost √ High energy efficiency	√ Very low emission √ No range anxiety √ Higher fuel economy compared with ICE	√ Very low emission √ No range anxiety √ Higher fuel economy compared with ICE
Major issues	√ Battery sizing, power density, energy density, management, lifetime, and cost √ Charging facilities √ AI driving.	√ Optimal control and management between multiple energy sources √ Battery sizing and management	√ Fuel cell life cycle, cost and reliability √ Hydrogen production and distribution infrastructure

According to the level of electric power and the function of the electric motor, HEVs can be classified into the following categories.

1.3.3. Mild Hybrid

It not only has more than two energy sources but also has two power sources to drive the vehicle at the same time (for example, an internal combustion engine with an electric motor). In general, the internal combustion engine is a primary power source associated with an integrated starter generator (ISG) [26] as a secondary power source. ISG is installed between an internal combustion engine and transmission for auxiliary power. If the internal combustion engine needs to be started, the motor acts as a starter;; if primary power is insufficient during driving, ISG is used as an electric motor to aid the engine. The electric motor provides greater efficiency by replacing the starter and alternator with a single device that assists the powertrain. Mild hybrids generally cannot provide all-electric propulsion. For fuel economy improvement, the mild hybrid is estimated to be in the range of 10%-20%.

1.3.4. Full Hybrid

It, sometimes also called a strong hybrid, is a vehicle that can run on just the engine, a motor, or a combination. A large, high-capacity battery provides battery-only operation. The full hybrid vehicles have a split power path that allows more flexibility in the drivetrain by inter-converting mechanical and electrical power. To balance the forces from each portion, the vehicles use a differential-style linkage between the engine and motor connected to the head end of the gearbox. When such a vehicle uses this fully electric system, it becomes a "zero-emission vehicle" (ZEV). For fuel economy improvement, the full hybrid was estimated to be in the range of 20%-50%.

1.3.5. PHEV

It charges batteries from a wall outlet or in the garage, which improves the efficiency of utility power usage because the charging is done at night [26 - 28]. In addition, they can use grid to increase the spin reserve of peak power by bidirectional power transfer such as vehicle to grid (V2G) [29, 30] or vehicle to home (V2H) [31, 32]. Due to this reason, they can run solely on electric power, resulting in better miles per gallon (MPG) [13]. Toyota Prius PHEV, Mitsubishi Outlander PHEV, Nissan Leaf, Citroen Numero 9, and more manufacturers' PHEVs are on the market. For fuel economy improvement, the plug-in hybrid can be improved by 100% if the ICE is not used (*e.g*., in urban drive cycles). Increasing the battery size allows ZEV operation for short trips and thus results in an important reduction in fuel consumption and greenhouse gas emissions. Degrees of hybridization are shown in Table **1.2**.

Table 1.2. Degrees of Hybridization.

Capability	**Micro Hybrid**	**Mild Hybrid**	**Full Hybrid**	**Plug-in Hybrid**
Automatically stops/starts the engine for reducing emission	yes	yes	yes	yes
Regenerative braking and excess energy during coasting.	-	yes	yes	yes
Uses an electric motor to assist a combustion engine	-	yes	yes	yes
Can drive vehicle using only an electric motor	-	-	yes	yes
Plug-in charge for extended all-electric range	-	-	-	yes

1.4. A Brief Review of EVs Traction Motors

The EVs traction motors for the propulsion system play a very important role that considers acceleration, deceleration, acoustic noise, efficiency, and cost. The position of EVs traction motors is divided into concentrated [33] and distributed [34 –36]. Table **1.3** provides a summary of some of the salient examples in recent years for light-duty vehicles [36 – 38], most commonly used, including Permanent Magnet synchronous motor (PM) and Induction Motor (IM). The IM has the characteristics of easy manufacturing and low cost, but its disadvantages are that it has lower efficiency and lower power density. The PM has the characteristics of high power density and high efficiency. However, it is expensive due to the use of rare earth material. DC motors produce torque and speed with a mechanical commutator (coal brush), which is not the same as PM motor using an electric commutator. Because their torque vs. speed characteristics suit the traction need well and their speed controls are simple, it has become a mature technology in electric propulsion. However, dc motor drives have low efficiency and a higher need for maintenance. The French automaker PSA Peugeot Citroën introduced the HEV version of the well-known Berlingo (1998-2005), which is called Dynavolt, with a dc motor as electric propulsion [34]. The Switched Reluctance Motor (SRM) is gaining more attention because it has a simple and rugged construction and fault-tolerant operation. Some of its disadvantages are acoustic noise, torque ripple, and electromagnetic interference (EMI). The performance evaluations compared with DC motor, PM motor, IM motor, and SRM motor for application on light-duty vehicles are shown in Figs. (**1** to **9**) [34, 37]. Based on the main characteristics of the electric propulsion, each of them is graded from one to five points, where five points mean the best, according to which IM motor is the most adapted candidate for the electric propulsion.

Table 1.3. Electric Motors in Commercial EVs [36–38].

Vehicle	Market Release	Vehicle type	Motor Type	Bus Voltage [V]	Peak Power [kW]	Peak Torque [Nm]	Max. Speed [rpm]
VW bora compact sedan	2019	BEV	PMSM	400	75	1000	1600
KIA soul	2018	BEV	PMSM	375	81.4	285	8000
Regera hyper-car	2018	HEV	PMSM	800	199	792	3250
BMW i8	2018	HEV	PMSM	340	105	250	-
VW e-Golf	2017	BEV	PMSM	323	100	290	12000
Tesla Model S	2012	BEV	IM	-	441	285	16000
Tesla Model X	2016	BEV	IM	-	193(front) 375(Rear)	249(front) 649(Rear)	6100(front) 5950(Rear)
Tesla Model 3	2017	BEV	PMSM	-	192	430	-
Hyundai Ioniq	2017	HEV	PMSM	-	32	169	-
	2018	EV	PMSM	-	88	295	-
	2018	PHEV	PMSM	-	45	169	-
Hyundai Sonata	2017	PHEV	PMSM	-	50	205	3300
Toyota Prius	2016	PHEV	PMSM	-	53	163	-
Toyota RAV4	2014	PHEV	IM	-	115	370	-

Fast and quick torque response is the main characteristic of high dynamic performance traction drive. The ideal torque-speed profile for traction applications is the constant power in all the speed ranges [39]. A well-controlled electric motor drive has a torque-speed profile close to the ideal, as shown in Figs. (**1.10** to **1.13**)

[40, 41]. In these figures, a well-controlled electric motor drive includes two distinguishable operating regions: constant torque and constant power region. The maximum torque in the constant torque region is determined through the inverter current rating, while the constant power region is present due to inverter voltage and current limits [42]. In the last region (flux weakening), both torque and power decay due to the increasing influence of back electromotive force (Back-EMF) [41]. The extended constant power operation is important for both vehicle initial acceleration and coasting travel and requires less acceleration power [43].

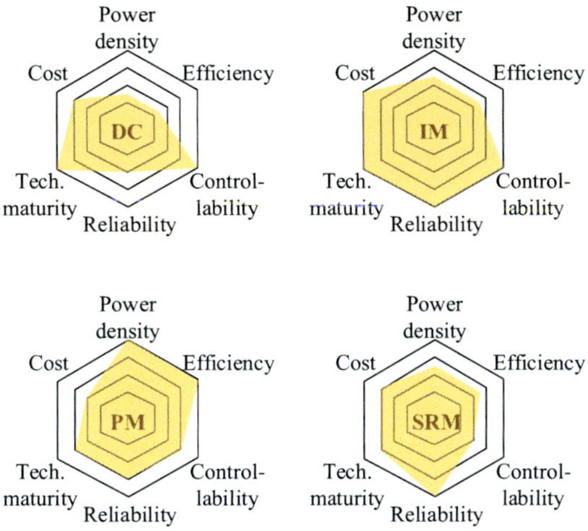

Fig. (1.9). Various features of electrical motors for EV and HEV application [34, 37].

Field-oriented control (FOC) and direct torque control (DTC) [44] are the most popular methods for AC motor drives, such as IM, PM, *etc.*. They were employed to control motor speed or torque according to reference command. FOC control scheme (Space vector control) can be derived based on the d-q model of AC motors, which achieves the decoupling of torque control and field control; similar characteristics such as the dc motor can be obtained. DTC has no need for complicated coordinate transformation. The torque is controlled directly by controlling the angle of the stator flux linkage vector, which provides a very quick response. Both FOC and DTC can generate PWM by adopting the Space vector pulse width modulation (SVPWM) technique [45, 46].

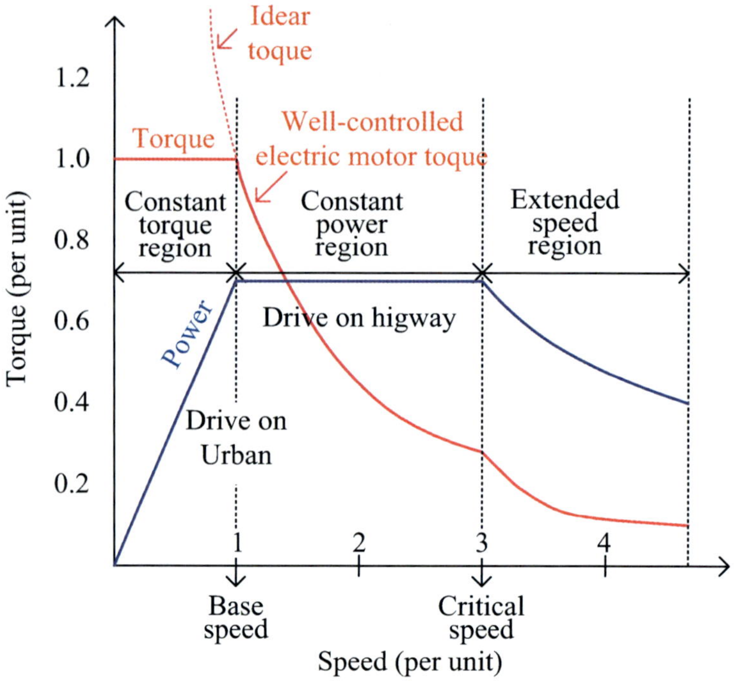

Fig. (1.10). Typical speed-torque characteristics of EVs & HEVs from the electric traction motor.

The EV electric motor makes numerous stops and starts on the road, which should satisfy several essential characteristics that are as follows: 1) high torque at low speed to satisfy hill-climbing conditions; 2) high-efficient range during a wide range of speed; 3) overload capability, typically twice the rated motor torque for a short period; 4) fault-tolerant capability; 5) high-power density according to weight and volume; 6) quick and fast dynamic response for driver needs; 7) high ruggedness in harsh environments; 8) low acoustic noise; 9) high-power and high speed on highway driving; 10) small size, reduced weight, the lower moment of inertia; 11) acceptable cost. More details of the operation characteristics of these motors are described below.

1.4.1. DC Motor

The DC motor is a classical electromagnetism application according to Ampère's circuital law. A brief description of the dc motor and driver is given in Fig. (**1.11**). The stator of the motor with two poles is a permanent magnet; the rotor of the motor is a coil with a rotating slip ring through a coal brush. The controller regulates i_a by Pulse Width Modulation (PWM) technology to turn on-off switching S1~S4. When the motor runs clockwise, S1 and S4 = PWM and S2 and

S3 = off; when the motor runs counterclockwise, S1 and S4 = off and S2 and S3 = PWM. The i_a injects the coil *via* the coal brush, which produces torque by electromagnetism. The angle of the coal brush relative to the permanent magnet of the stator is fixed, which makes the rotor run at optimal efficiency.

Conventionally, the dc motor with coal brush is widely used due to its excellent torque-speed characteristics and simple speed control. However, low efficiency, low power density, and additional maintenance requirements, such as replacing the brush, make it less attractive in EV applications. However, in lower power propulsion applications, the brushed dc motor system is still applied because of its simplicity, low cost, and mature technicality.

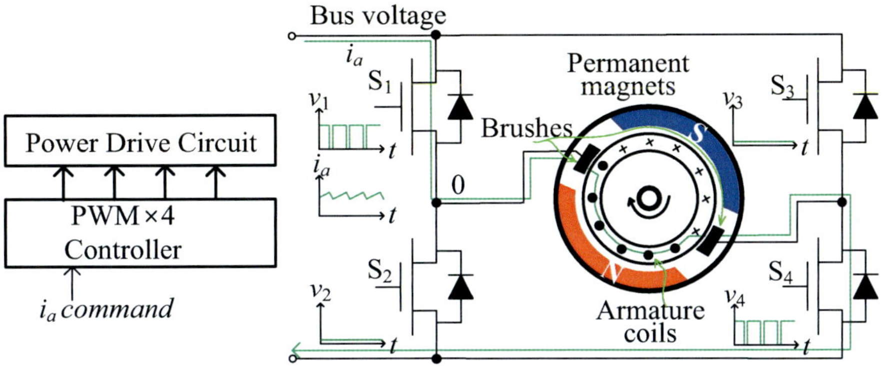

Fig. (1.11). Schematic representation of dc motor with driver.

1.4.2. Permanent Magnet Synchronous Motor (PM)

Due to permanent-magnet excitation, PM brushless machine is inherently efficient, high power density, and reduced weight and volume gave power rating [47]. Permanent Magnet Synchronous Motors (PMSM) use a set hall sensor to provide feedback on rotor position to the driver, allowing for alternate current follow rotor position. A brief description of the PMSM motor with hall sensor to feedback rotor position and driver is given in Fig. (**1.12**). The stator has a three-phase winding; the rotor is a permanent magnet with two poles. The controller regulates three-phase current (i_a, i_b, and i_c) according to the detection of the rotor position to turn on-off switching S1~S6 using Pulse Width Modulation (PWM) technology, which has either sinusoidal or trapezoidal waveforms. The 3-phase current produces a spin magnetic field to synchronize rotor speed to produce torque.

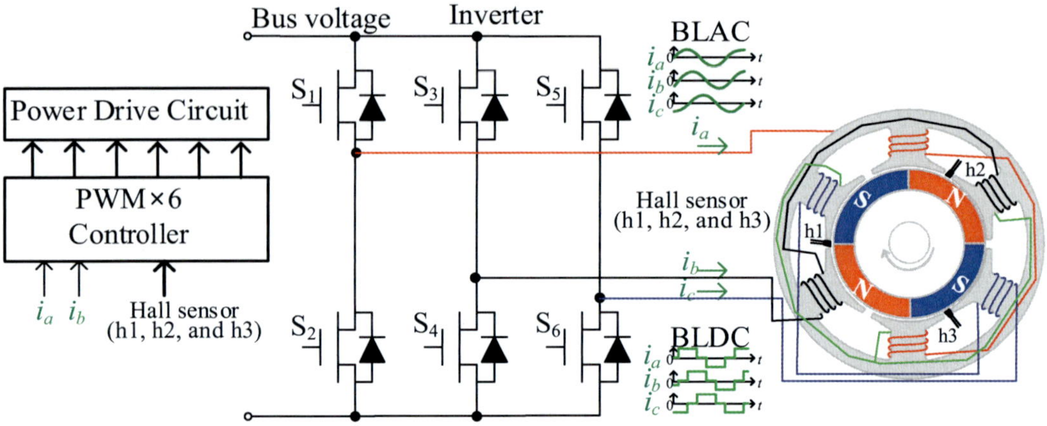

Fig. (1.12). Schematic representation of PM brushless motor with driver.

A motor runs on the three-phase winding feedback sinusoidal or trapezoidal back-EMF corresponding brushless AC (BLAC) motor or brushless DC (BLDC) motor. Fig. (**1.12**) shows the schematic representation of a typical PM drive. In both BLAC and BLDC drives, rotor position information is necessary, which generally employ low-cost Hall sensors or/and a relatively high-resolution encoder to detect rotor position. However, numerous sensor-less techniques have recently been developed or are under development for both BLDC and BLAC drives. The PM motors inherently have a short constant-power region (Fig. **1.10**) due to flux weakening capability in the high-speed range. To extend the speed range of PM motors, the conduction angle control is employed [48].

There are several types of PM motors, *i.e.*, surface-mounted PM (SPM) motor, interior PM (IPM) motor, flux switching PM (FSPM) motor, and so on, for EV drive [19, 49 - 51]. FSPM motor feedback sinusoidal EMF is more suitable for BLAC control and it also has small torque ripple and high-speed applications. However, the FSPM has disadvantages such as low magnet utilization ratio and expensive manufacturing cost than IPM [19, 51]. SPM motor for traction is easy to manufacture, which usually has concentrated stator coils and shorter end connections which leads to lower copper loss. The armature-reaction field is small, the inductances of the d-axis and q-axis are almost the same, and the reluctance torque is nearly zero [19, 41]. As for the losses, the SPM motor suffers from PM loss and joule loss due to the de-excitation current at high-speed. The PM loss can be reduced by segmentation in both circumferential and axial directions [49]. IPM motor with the V-shaped PMs has a strong structure, high-speed operational capability, and magnetic saliency effect [51 – 53]. For the magnetic saliency effect, it can lead to the reluctance torque generation; therefore, the high-speed operation of a motor through flux weakening control is enabled

[53, 54]. Therefore, the IPM motor with the V-shaped PMs has a much larger overload torque over the speed range, a safer back electromotive force (EMF) in uncontrolled generator operation [55], and little sensitivity to PM temperature. The IPM motor has higher joule loss at low speed, but a proper stator slot number and rotor segments can keep the harmonic loss under control. Particularly, IPM motor is widely used in EVs, such as Toyota Prius with a V-shaped rotor.

1.4.3. Induction Motor (IM)

The IMs motor is widely accepted as the most potential candidate for electric vehicle propulsion due to its mature technology, ruggedness, reliability, low maintenance, low cost, and ability to operate in rigorous environments [11, 41]. The IM motor is already applied in many light-duty vehicle EVs including the Tesla Model S, Tesla Model X, Toyota RAV4, *etc.* A brief description of the PM motor and driver is given in Fig. (**1.13**). In this figure, the controller uses FOC or DTC algorithm to generate three sinusoidal waveforms with a phase shift 120□ difference through close-loop-control with i_a and i_b signal to drive the IM motor. Both a high starting torque and a low starting current can be achieved since the supply voltage and frequency are variable. By appropriate choice of supply voltage and frequency, the starting torque can be almost as high as the maximum torque, while a high efficiency can be achieved by minimum slip control [56, 57].

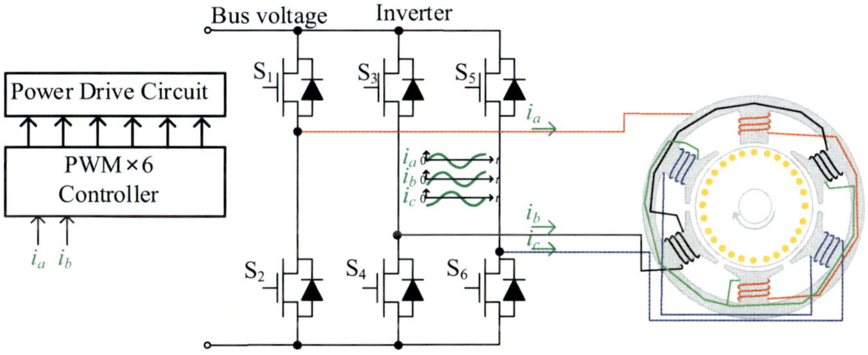

Fig. (1.13). Schematic representation of IM motor with driver.

For conventional IMs, the constant power range typically extends to 2–3 times the base speed. However, for traction machines with flux-weakening control to extend speed range, this can be extended to 4–5 times the base speed [41, 58]. Without using any PMs, the manufacturing process is significantly reduced as well as the overall cost. Optimal flux, maximum efficiency, and minimum acoustic noise are also discussed [41].

1.4.4. Switched Reluctance Motor (SRM)

Switched reluctance motors (SRMs) are characterized by rare-earth-free and wide-speed-range advantages and are becoming an attractive technology for EV/HEV applications [59, 60]. RMs' main characteristic is the use of rotor salient poles. The very robust rotor is cheap to produce and not temperature sensitive [61, 62]. Its efficiency remains high over a wide speed range. Efficiency values over 95% have been reported [63]. The high rotor inductance ratio makes sensor-less control easier to implement [63, 64]. The high ripple torque resulting in higher noise and vibrations is the main drawback. The investigations considered the stator/rotor geometry, power/torque density, reduced noise, and vibration [41, 65, 67]. A fault-tolerant topology system with advanced control functions is subjected to fault conditions, including open-circuit and short-circuit faults; the proposed converter starts its fault-diagnosis procedure to locate the fault [68].

A drive with a 6-phase/8-pole SRM is illustrated in Fig. (**1.14**). In this figure, the controller regulates three-phase unipolar current i_a, i_b, and i_c by Pulse Width Modulation (PWM) technology to turn on-off switching S1~S6, and SRM motor rotates clockwise with pulsated torque.

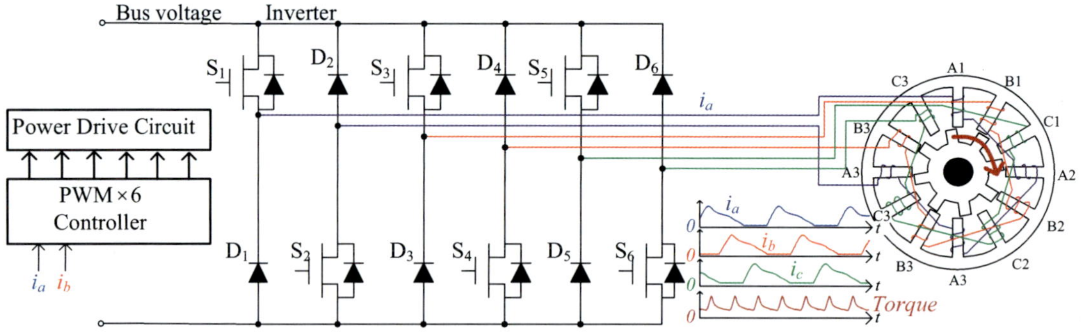

Fig. (1.14). Basic winding structure and driving topology of a 6-phase 12/8-pole SRM.

The efficiency of electric motors depends on the working points. Their characteristic footprints are shown in Fig. (**1.15**). If these motors with the same peak efficiency are compared, PM motors are more efficient in overload transients at a constant speed, whereas RM has better performances at high-speed overloads [69].

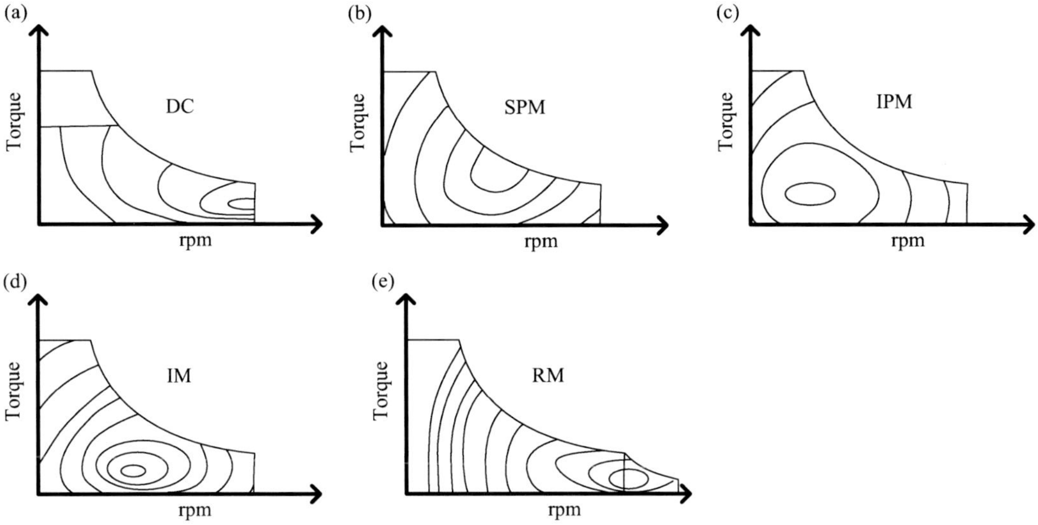

Fig. (1.15). Efficiency map for (a) DC motor, (b) surface-mounted PM motor, (c) internal-mounted PM motor, (d) IM motor, and (e) RM motor.

1.5. Li-ion Batteries

In an EV, the battery pack is not easy to design because of the following requirements: high power density and energy density, long cycle lifetime, short charging time, affordable cost, and reliability, which should be achieved simultaneously. Li-ion batteries are recognized as the most competitive solution to be used in EVs. The advantages of Li-ion batteries are high energy density, long life-cycle, heat and cold resistance, safety, and commercialization [70, 71]. The Li-ion batteries or energy storage system is the power source of the vehicle. It can be repeatedly charged and discharged, so it is called a secondary cell. The high energy density (compared to lead-acid, nickel-metal hydride batteries), high power density, and fast charging characteristics are the main research direction of vehicle battery technology. These technological developments and advancements are crucial for the ICE to become a historical achievement.

The storage system usually uses 20 to 40 cells as a group in parallel and uses 96 to 120 as a group in series. The batteries in series provide higher voltages; batteries in parallel provide larger currents. In addition, to avoid excessive discharging batteries, these batteries provide sufficient dynamic power to the motor to meet driver needs.

The main materials of a lithium battery include a positive electrode material, a negative electrode material, an electrolyte, and a separator. Energy storage and release rely on lithium ions moving between the positive and negative electrodes.

The battery is charged by an external electric field. The electrons originally in the positive electrode material move to the negative electrode through the charger. Lithium ions are released from the positive electrode lattice, pass through the separator, and finally into the negative electrode lattice, as shown in Fig. **(1.16)**.

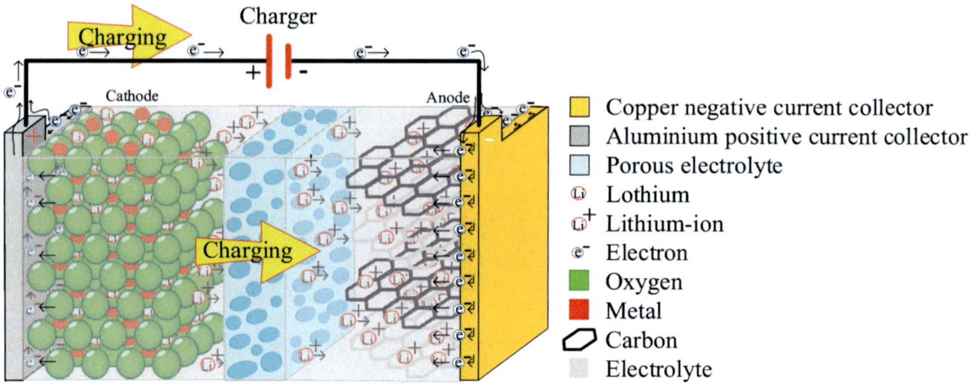

Fig. (1.16). Lithium-ion battery chemistry during charging.

When the battery is discharged, the electrons work on the load and move to the positive electrode; lithium ions are released from the negative electrode lattice, pass through the separator, and finally embedded in the positive electrode lattice, and combine with the electrons, as shown in Fig. **(1.17)**. The rate of moving lithium ions determines the battery power density.

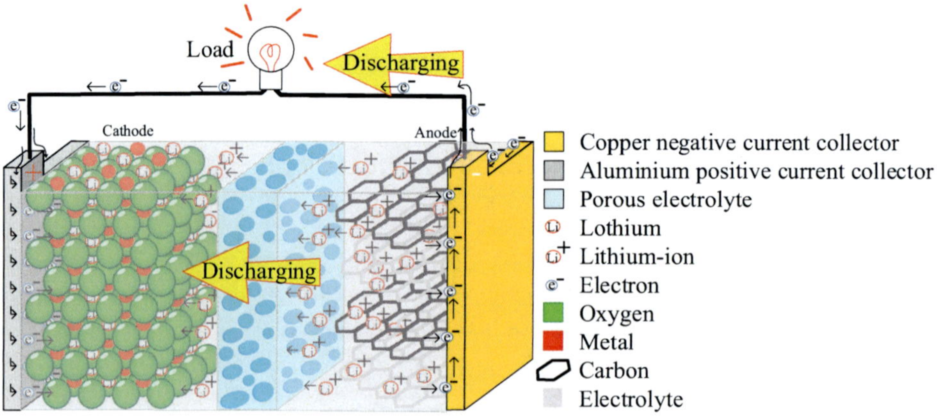

Fig. (1.17). Lithium-ion battery chemistry during discharging.

These Li-ion batteries are a whole lot better than Nickel-Metal Hydride (NiMH) cells and lead acid cells found in EVs of yore, but they too have their limitations. One of the most difficult challenges in battery design is increasing energy density while also maximizing battery life span [72, 73]. Li-ion batteries have achieved better combinations of these parameters than any previous technology. Yet there is still a tradeoff between energy and life, even within the family of Li-ion.

The Li-ion battery has good charging and discharging electrical characteristics. Its charging and discharging curves are illustrated in Fig. (**1.18**). While charging, the charging capacity increases gradually, with the charge voltage maintaining a constant current between t0 and t1. When the voltage reaches a maximum, the current decreases exponentially. On the other hand, the capacity discharge maintains an almost constant voltage and current to the load, although there is a small decrease and increase in the voltage and current values, respectively, until the cell capacity reaches the minimum acceptable level, which is set by the manufacturer. For example, the end of discharge voltage is 3.0 V, as shown in Fig. (**1.18 B**).

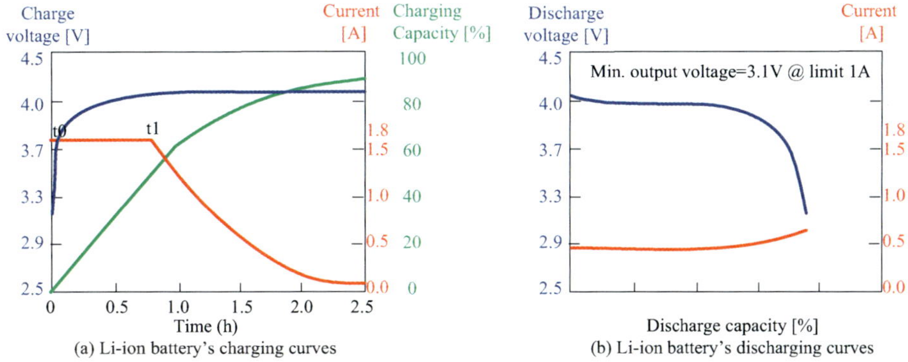

Fig. (1.18). Typical characteristics of the lithium-ion battery.

However, discharging capacity is less than the charging capacity. The state of the charge (SOC) is determined by these leakage energy causes, which include the charging and discharging cycle, ageing, operational environment, operating temperature, and so on. The SOH indicates how much the battery has been abused, and its performance has been degraded [74, 75].

There are two kinds of aging: aging from use, called "cycle life," and aging with time, called "calendar life" [76, 77]. Manufacturers define cycle life for Li-ion cells as the number of full discharge-charge cycles required to reduce a cell's capacity to a fraction of its original state (A common threshold used in the laptop industry is 80 percent).

Other negative effects of cycle life are as follows:

- Stay away from very low and very high states when discharging and charging. Voltages below 3.00V/cell and over 4.15V/cell cause more stress on the battery.
- Stay away from very high charge rates; charging at a rate greater than C/2 reduces the life of the cell.
- Stay away from charging at temperatures below 0° C.
- Stay away from very high discharge rates.

Calendar Life Li-ion cells lose capacity with time, even if they are just sitting on a shelf. They lose the most capacity early in their life and then continue to lose capacity gradually thereafter. Some factors that shorten the calendar life are as follows:

- Life-time average temperature and time spent at high states of charge. Batteries would last the longest if they were stored in a refrigerator at a very low state of charge.
- They are the fastest when stored in a hot place at a full state of charge.

A liquid cooling system is used by some manufacturers of important inventions to improve battery lifetime. This technology keeps the batteries at a favourable temperature even under extreme ambient temperature.Table **1.4** shows the comparison between Li-ion batteries with other types of batteries.

Table 1.4. Characteristics of li-ion batteries compared with other types of batteries

Battery Type	**Li-ion [78]**	**Lead acid [79]**	**Ni-Cd [80]**	**Ni-MH [81]**	**Zn-Br [82]**	**Fe-Cr [83]**
Power Density [W/kg]	1800	180	150	250-1000	-	70-100
Energy Density [Wh/kg]	110-160	30-50	45-80	60-120	35-54	20-35
Nominal Volt. [V]	3.6	2	1.25	1.25	1.67	1.18
Operating temp. [°C]	-20~60	-20~60	-40~60	-20~60	-20~60	-40~60
Cycle Live	500-1000	200~300	1500	300~500	>2000	-
Charge Efficiency [%]	95~	79	-	-	-	-
Energy efficiency [%]	80	70	60-90	75	80	66
Overcharge Tolerance	Very low	High	Moderate	Low	High	Moderate
Self-discharge	Very low	Low	Moderate	High	Low	High
Thermal Stability	Most Stable	Least Stable	Least Stable	Least Stable	Least Stable	Stable

1.6. EVs on a Power Distribution System

BEV or PHEV charging batteries connect a wire to a grid. Therefore, BEVs and PHEVs will become a distributed power source to be a commitment by an aggregator [84, 85]. They aid the power grid to provide efficiency, stability, and reliability. In contrast, they can put the grid in a critical condition when heavily charging batteries during peak power demand.

The power demand of a certain area in winter [86] is shown in Fig. (**1.19**). It has a wave trough between 0h-6h which is the period of minimum power demand; a wave crest between 18h-20h is the period of the highest power demand. Regardless of the demand, the amount of power generated in the grid must be equal to the demand to meet the voltage and frequency of grid stability. Therefore, the grid is a large servo system, which keeps the balance between output power and demand at all times.

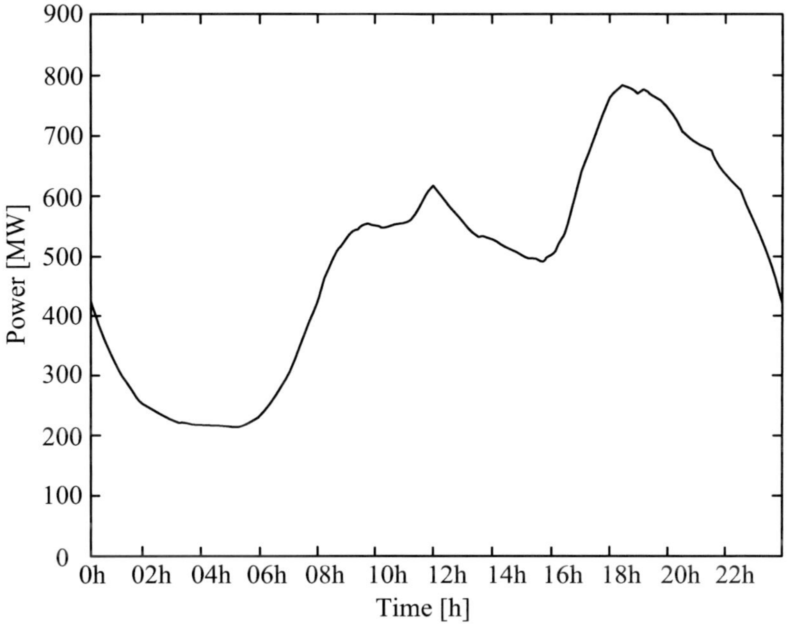

Fig. (1.19). The power demand of a certain area in winter.

The maximum power generation and the proportion of generation of units are shown in Fig. (**1.20**). In this figure, "Nuclear: 20, 0.038" means that nuclear power generation accounts for 20%, and the cost is \$0.038/kWh. The cost determines the order of units. For example, baseload power such as nuclear units and coal units provide a lower generated cost. The higher generated cost, such as oil units, is only turned on during peak power demand. Nuclear units and coal units are baseload power units that are continuously generated to meet most of the demand. The hydro units are a renewable energy source that acts as a load or generator to provide bidirectional power transfer. They pump water to the reservoir during 0-6h and generate power to the power grid during 6-24h. Therefore, the main function of the hydro units aids the power grid in economic operation.

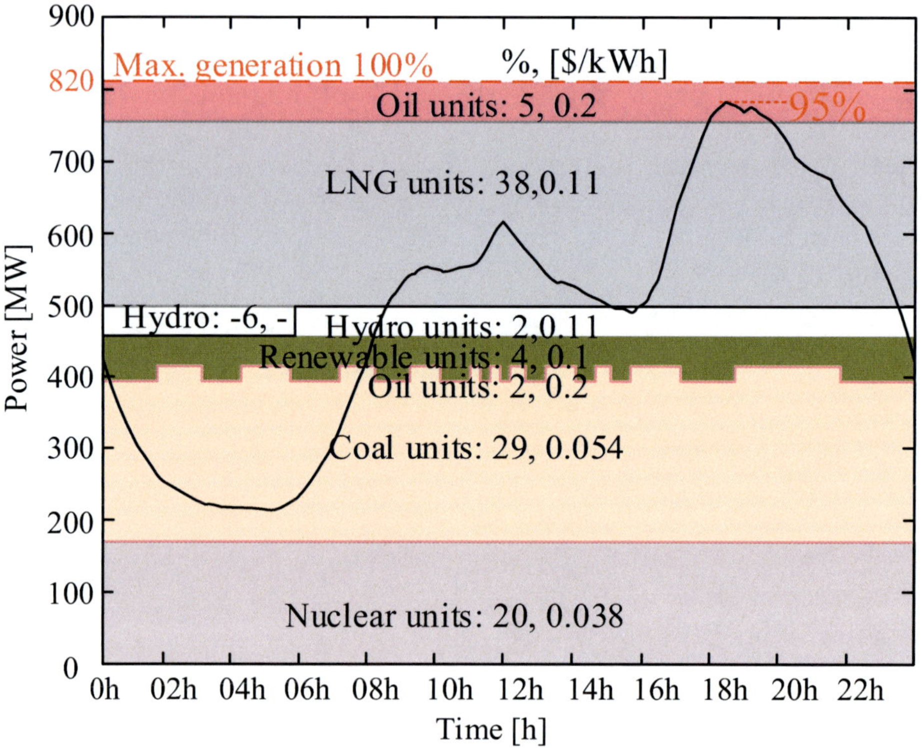

Fig. (1.20). Proportion of generated power by each generator.

The Liquefied Natural Gas (LNG) generates more power than coal units because of less emission. However, the cost ($0.11/kWh) of LNG is more than coal units (0.054/kWh). Oil units can immediately start generating power, allowing them to satisfy surge demand such as peak power demand (18 h-20 h). However, its cost of generation is the most expensive, which is 5 times that of nuclear units. The proportion of power generated by oil units is only 5%. The renewable units generate only 4% power, which provides intermittent power to the grid. In this situation, the power grid absorbs the impacts of intermittent power by coal units. In Fig. (**1.20**), the spinning reserve is only 5% in 18h-20h. Therefore, the situation of this power grid system is critical.

The number of EVs is expected to grow rapidly in the coming years. An EV battery charging rate is slow, *i.e.*, 1.5 kW [5] or fast, *i.e.*, 400 kW [87]. However, the uncoordinated charging of these vehicles can put severe stress on the power grid. EVs are capable of charging their batteries to 80% capacity. Slow-rate charging takes 10 hours and fast-rate charging takes half an hour. On the grid, the danger arises when huge numbers of EV batteries using fast charging rates are used without being managed. For example, 200 EVs use fast-charging batteries to add 80 MW demand on the grid during peak power demand. If the total demand is more than max. generation at 19 h, then there is an output power cut-off on a certain area to avoid generators' breakdown, as shown in Fig. (**1.21**).

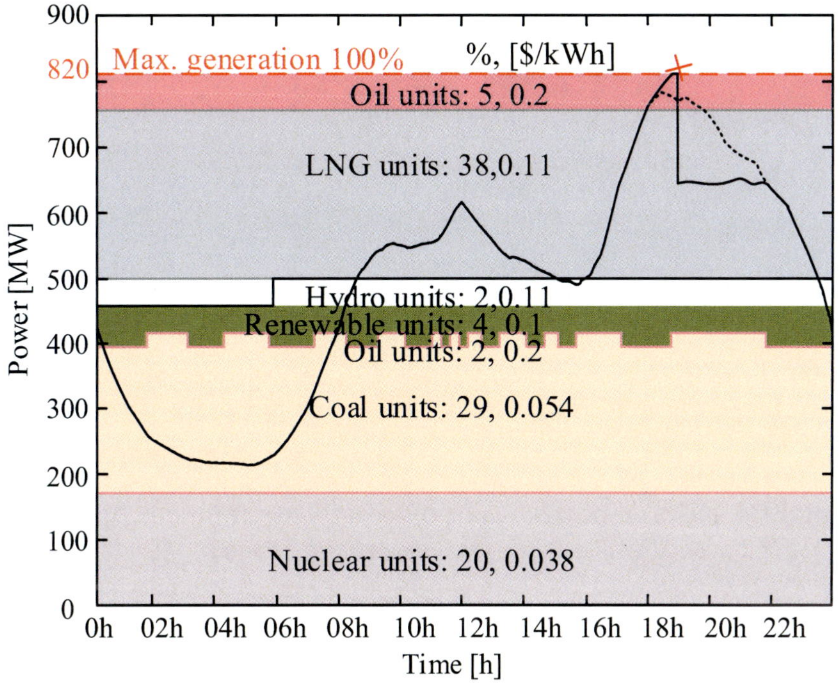

Fig. (1.21). The power grid de-rating output power during peak power demand.

As above, the EVs must be managed to avoid power grid breakdown or de-rating. Aggregator aggregates the contracted EVs and contracted factories to be a virtual power plant (VPP) [88 - 90]. The VPP de-load during 18h-20h as shown in Fig. (**1.22**). In the Figure, the optimal day-ahead market bidding strategy of an aggregator of a power distribution system controls the spinning reserve up to 10% and shifting the peak demand to 02h-07h. Therefore, the aggregator creates a win-win situation for both power plants and contracted EVs and factories, which reduces the number of power plants and brings benefits to the power distribution system.

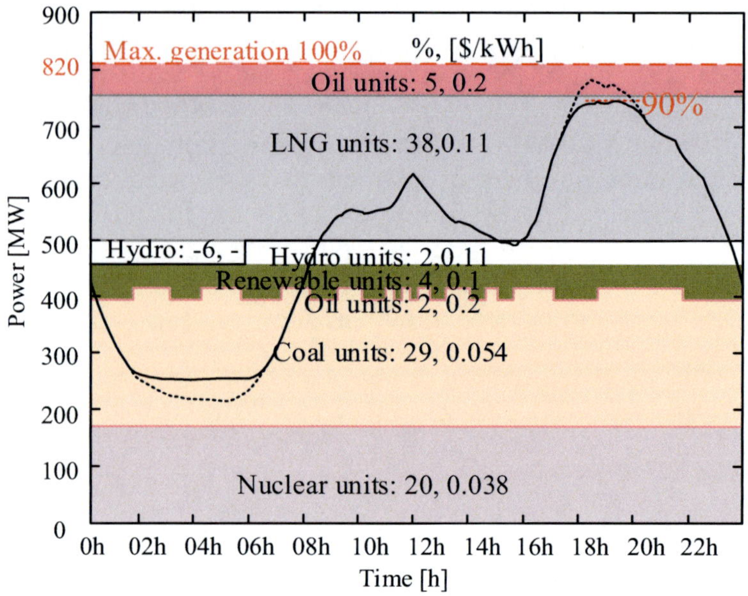

Fig. (1.22). Contracted power de-load and spinning reserve up to 10%.

Vehicle-to-grid (V2G) has been used as the bidirectional power transfer between the EVs and the power grid traditionally. However, recently, these EVs are more likely to present a general concept of EVs providing a buffer to store energy. In this aspect, the benefits of V2G applications have been widely studied, including load shifting [91], reduced power loss, power unit commitment, and the impact of charging PHEV or BEV on distribution power systems [92 – 95].

A city with a power distribution system is illustrated in Fig. (**1.23**). In this figure, the power distribution system includes a centralized power plant, distributed energy storage units, independent system operator (ISO), aggregator, and power transmission network. The optimal day-ahead market bidding strategy of an aggregator of a power distribution system accepts direct assignment of ISO which provides economical operation, reduces emission, optimal unit commitment, and stable voltage and frequency on the power grid.

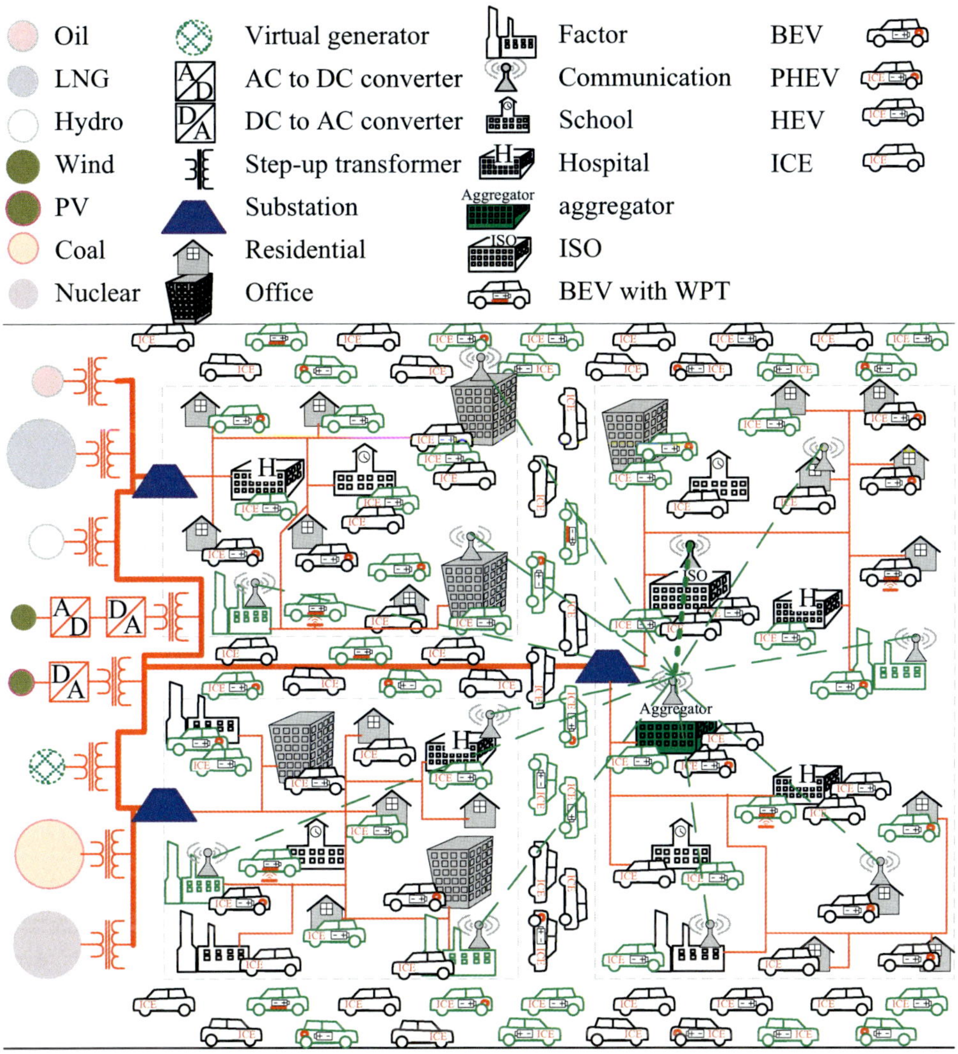

Fig. (1.23). A power distribution system with an aggregator in a city.

There is bidirectional power transfer by EVs, as shown in Fig. (**1.24**). In this figure, the EVs charging batteries during 01-07h (G2V) and EVs output energy to the grid during 18h-20h (V2G) are depicted. In addition, the spinning reserve of the power grid is up to 10%.

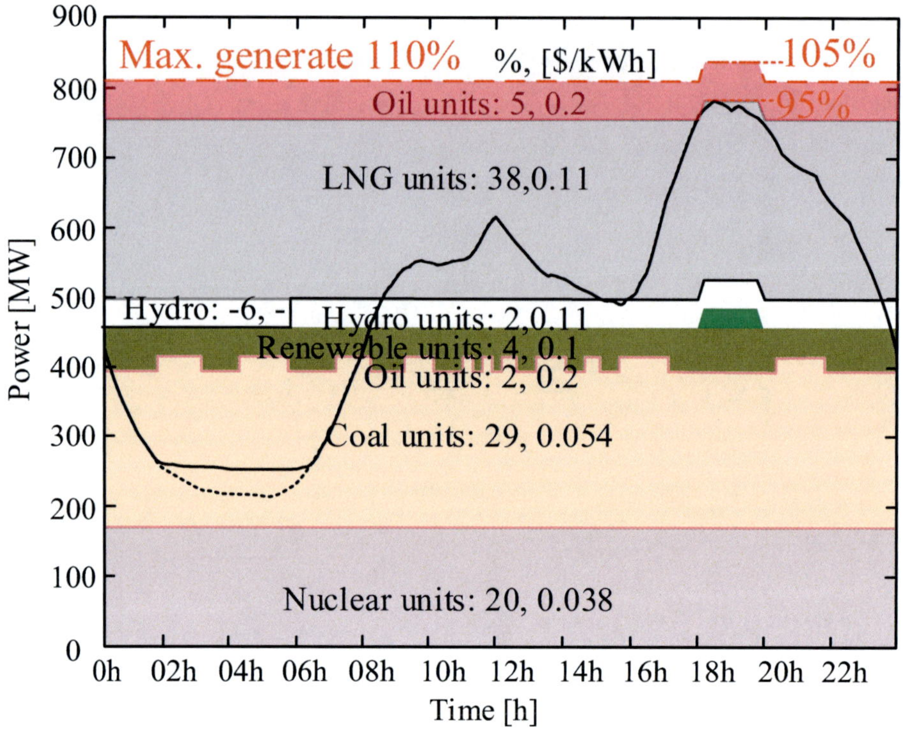

Fig. (1.24). Bidirectional power transmission by EVs keeps the power grid running efficiently and increases the spinning reserve by up to 10%. Batteries charging for BEV and PHEV.

Between 2030 and 2050, the worldwide ICE vehicle market will be closed. It will be replaced by EVs or HEVs, which have zero or low emissions and high-efficiency performance. Before the battery runs out of energy, they use traditional conducive charging or novel wireless charging to charge the batteries between the power grid and the vehicle. As compared to ICEs, EVs require more time for refueling, dedicated connectors, and fewer charging stations.

1.6.1. Conductive Battery Charging

The EVs or PHEV receive energy from the power grid by a physical plugging-in connector and store the energy into a battery pack. The charging rate is determined to input power by the capacity of the charger and batteries; it can be charged and discharged many times over its useful service life.

The conductive charger consists of an ac–dc rectifier and dc-dc converters or a low-frequency ac to high-frequency ac converter with PFC. They are classified as on-board and off-board chargers. In on-board chargers, rectifiers and battery current regulators reside inside the vehicle, whereas rectifiers and battery regulators reside outside the vehicle for off-board chargers. They are classified based on the power transfer level. Level 1 and Level 2 are the on-board chargers that should be located on the vehicle and the connector compliance with SEA J1722 is required [96 - 99]. The level 1 charger has a power output of 2 kW (120V/15A), whereas the level 2 charger has a power output of 4–20 kW (208V, 240V/up to 80A). Level 3 dc fast chargers, sometimes known as superchargers, have a power range of 50–150 kW, allowing them to charge a battery in less than an hour. It can be installed in highway rest areas and residential areas as a refueling point.

The state-of-the-art dc fast chargers convert the three-phase ac voltage up to 480 V or higher into the desired dc voltage by two power electronics-conversion stages: ac to dc rectification stage with power factor correction (PFC), which converts three-phase input ac voltage into an intermediate dc voltage, and a dc to dc stage, which converts the intermediate dc voltage into a regulated dc voltage. The dc voltage source injects energy into the off-board charger of the vehicle *via* a wire cable. Table **1.5** shows a comparison between various models of EVs and PHEVs [76, 99]. Both level 2 and dc fast charging block diagrams are shown in Fig. (**1.25**). Due to the on-board charger (level 1 or level 2) having a relatively low-power rating, it is suitable for overnight charging. The level 3 or dc fast charging charge battery in less than 1 hour to decrease long time charging. The models of dc fast charging include J1772 combo, Tesla supercharger, CHAdeMO, CCS, *etc.* [87, 97, 100].

Table 1.5. Various Ev Types And Phev Models with their Charging Level and Battery Capacity [76, 99].

Mode	Driving Distance/ 1 Hour of charging	Battery Energy Size [kWh]	EV Type	Type of Charging	Max Charge Time [Hr]	Max Charge rate [kW]
Porsche Panamera S E-Hybrid	10	9	PHEV	Level 1/ Level 2	10/2.5	3
Audi A3 e-tron	11	9	PHEV	Level 1/ Level 2	8/2.5	3.3
Cadillac ELR	11	16.5	PHEV	Level 1/ Level 2	12.5-18/5	3.3
Chevy Spark EV	11	19.44	100% Electric	Level 1/Level 2/ dc fast charging	20/7/0.33(80%)	60
Chevy Volt	11	18.4	PHEV	Level 1/Level 2	13/4.5	3.3
Ford C-Max Energy	11	7.6	PHEV	Level 2	2.5	3.3
Ford Fusion Energy	11	7.6	PHEV	Level 1/ Level 2	7/2.5	3.3

(Table 1.5) cont.....

Mode	**Driving Distance/ 1 Hour of charging**	**Battery Energy Size [kWh]**	**EV Type**	**Type of Charging**	**Max Charge Time [Hr]**	**Max Charge rate [kW]**
Mercedes S550 Plug-in Hybrid	11	8	PHEV	Level 1/Level 2	2/4-5	3.3
Mercedes S350 Plug-in Hybrid	11	6.2	PHEV	Level 2	2	3.3
Smart Electric Drive	11	17	100% Electric	Level 2	2.5	7.2
Toyota Prius Plug-in	11	9	PHEV	Level 1/ Level 2	5.5/2	3.3
Mitsubishi i-MiEV	11	16	100% Electric	Level 2 (16A, 10A, 8A)/ Level 3	6, 8, 10/0.5	3.6, 2.4, 1.9, 2/32
Nissan LEAF	11/22	24	100% Electric	Level 1/ Level 2/ Level 3	22/8/.5	3.3/6/48
Porsche Cayenne S E-Hybrid	12/24	11	PHEV	Level 2	3, 1.5	3.6, 7.2
Volkswagen e-Golf	12/24	36.6	100% Electric	Level 1/ Level 2/ dc fast charging	24/9/0.5(80%)	3.6/7.2/70
Ford Focus Electric	22	23	100% Electric	Level 1/ Level 2	20\3-4	6.6
Fiat 500e	22	24	100% Electric	Level 2	4	6.6
Kia Soul EV	22	27	100% Electric	Level 1/ Level 2/ Level 3	24/4-5/0.72	6.6
Honda Accord Plug-in Hybrid	22	7	PHEV	Level 2	3	6.6
Honda Clarity Electric	25	25	100% Electric	Level 1/ Level 2/ dc fast charging	19/3/.5(80%)	NA
BMW i3	25	24	100% Electric	Level 1/ Level 2/ dc fast charging	22/8/0.5	48
Mercedes B-Class Electric	29	28	100% Electric	Level 2	2	10
Tesla Model S	29/58/255	85	100% Electric	Level 1/ Level 2/ Supercharger	0.67 (80%)	150

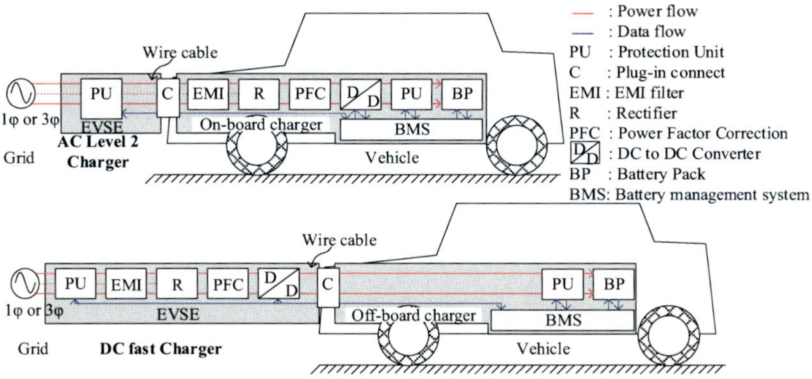

Fig. (1.25). Block diagram of the on-board charger and off-board charger.

1.6.2. Wireless Battery Charging

At present, wireless power transfer (WPT) for vehicle battery charging is an option in addition to conductive charging. The standardization of WPT by the SAE J2954 wireless charging task force is still at an early stage [101]. The WPT for vehicle battery charging without physical wiring provides convenience and safety [102]. Its schematic representation is shown in Fig. (**1.26**). In this figure, there are two units: the primary unit and the secondary unit. The low-frequency power energy from the grid to primary, which was converted into a high frequency (operation frequency) signal by a high-frequency converter (HFC) and injected LC tank (constructed by compensation network and coil). The LC tank amplifies this signal, radiates this signal, and is received by the LC tank of the secondary unit *via* an air gap. Finally, the received power will be absorbed by a load (battery pack).

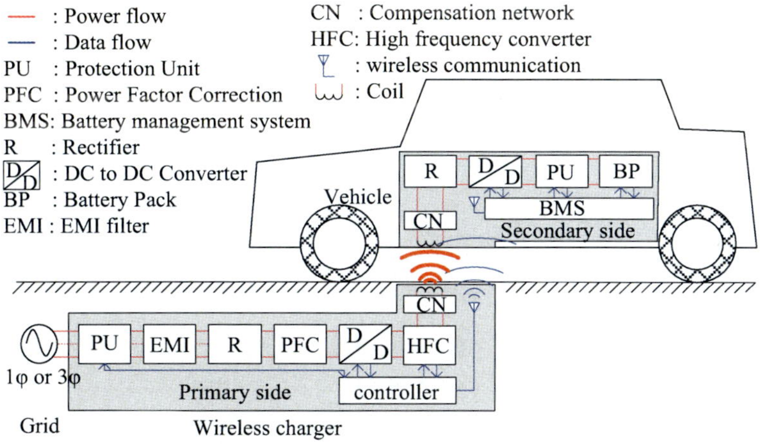

Fig. (1.26). Block diagram of wireless power charger with RF communication for the electric vehicle.

The resonant frequency of the LC tank is designed closed to operation frequency by tuning its characteristic impedance. Therefore, the characteristic impedance of the LC tank for the primary unit and secondary unit are the same. However, the primary side's characteristic impedance of the LC tank will be affected by the reflected load *via* mutual induction, making it difficult for the two sides' resonant frequency to be closed operation frequency simultaneously, especially for time-varying loads. There is a way to reduce the effective factor, which increases the distance between two LC tanks to reduce mutual induction even to zero. As mentioned above, resonant frequency of the LC tank for the primary unit and secondary closed to operation frequency was called magnetic resonances. MIT researchers presented a paper on power transfer using magnetic resonances, claiming that 60W of power can be transferred over a 2m distance [103]. The result surprised academia, and the WPT quickly became a hot research area. A lot of interesting works were accomplished with different kinds of innovative circuits, as well as the system analysis and control [104 – 106]. In order to transfer power more efficiently, the operation frequency is usually chosen at MHz level, and air-core coils are adopted. When the WPT is used in EV charging, the MHz operation frequency is hard to reach power and efficiency criteria because it is inefficient to convert a few hundred kilowatts of power at MHz frequency level using state-of-the-art power electronics devices. Moreover, air-core coils are too sensitive to the environment. If an air-core coil is attached to a vehicle, the magnetic flux will go inside the chassis, causing high eddy current loss as well as a significant charge in the coil characteristic impedance. To make the WPT system more practical in the LC tank, ferrite as a magnetic flux guide and aluminum plate as a shield are usually adopted [106]. The operation frequency of the LC tank with ferrite and aluminium is less than 100 kHz, which is magnetic field coupling between the transmitted and received LC tanks, which is similar to classic inductive power transmission (IPT) [108, 109].

The typical IPT technologies include induction cooking [110, 111], contact-less battery charging for electric vehicles [111 - 115], vehicle-to-grid systems [116 - 118], bidirectional inductive power transfer [119 - 121], mobile phones charging [122], medical implants [123, 124], dynamic charging on roadway applications [125, 126], avionic applications [127], power and data transfer [128 - 130], some special applications [131, 132], underwater vehicles [133, 134], mining systems [134], and cordless robots in automation production lines [136, 137].

The EV transfer power needs communication between the controller and the better management system (BMS). The transfer data include control signal, vehicle's ID, emergency message, charging voltage, current watt, and battery status, *etc.*, using RF equipment, as shown in Fig. (**1.26**). The wireless communication is quite mature and compliant with regulations of SAE J2931/1,

SAE J2836/6, and IEEE 802.11p, *etc.* [138]. However, this tutorial book introduces a novel wireless charging method that not only has massive power, compact size, and high transferring efficiency but provides two-frequency bandwidth and bidirectional communication using a data-attached mechanism without resorting to other RF devices. The novel communication provides charging and billing information and monitors safety during the charging period. Its diagram is shown in Fig. (**1.27**).

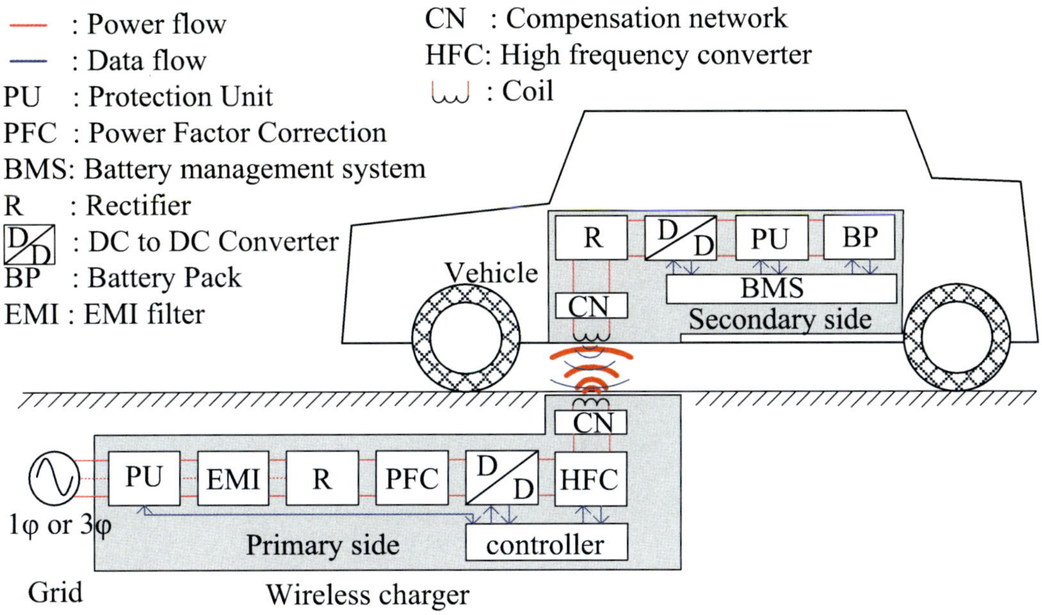

Fig. (1.27). Block diagram of coherent wireless power charging and data transfer for the electric vehicle.

CONCLUSION

Major advanced countries will ban fuel and diesel cars. Instead, hybrid electric vehicles (HEV), plug-in hybrid electric vehicles (PHEV), fuel cell electric vehicles (FCEV), and battery electric vehicles (BEV) will be used starting from 2030~2040.

This chapter widely reviews EVs' present technologies, including drivetrain configuration, electric motor drives, Li-ion battery technology, power distribution system management with charging/discharging of EVs, and conductive/wireless charger. The technologies discussed in this chapter will help readers to understand the global trend of EVs. They support efficient power, light-weight cars, which seems to be promising in the future, particularly in wireless/dynamic charging with on-the-road communication.

CHAPTER 2

The Trend of Green Power in the Civilian Vehicle Market

Abstract: In traditional centralized electricity generation units, fossil fuels, such as coal, oil, and liquefied natural gas, and steam turbines are used to generate continuous and stable power for loads. However, these units produce emissions and cause pollution. Green power, including wind and solar power, provides maximum environmental benefits and is associated with zero emissions and low pollution. However, climatic variations may result in the requirement of a high spinning reserve for a green power system; thus, the development of green power has been limited. An electric storage system (ESS) can overcome the aforementioned problem. The household vehicle market provides various types of vehicles, such as internal combustion engine (ICE) vehicles, electric vehicles (EVs), plug-in hybrid EVs, and fuel-cell EVs. However, many developed countries are planning to ban ICE vehicles between 2030 and 2040. EVs are driven with zero emissions by an electric motor. Moreover, these vehicles can achieve high torque with low speed and are thus more efficient than traditional ICE vehicles. However, the use of EVs is limited by their high battery costs and long charging time. With the progress of battery technology, battery and ESS costs are decreasing. Affordable electricity storage devices would enhance the efficiency of green power generation systems and the adoption of green power. Multiple power-generating resources are associated with low emissions; thus, vehicles running on traditional fossil fuels are gradually being replaced by vehicles running on these resources.

Keywords: Battery electric vehicle (BEV), Electric storage system (ESS), Electric vehicle (EV), Fuel-cell electric vehicle, Green power, Greenhouse effect, Household vehicle, Intermittent power, Internal combustion engine (ICE), Liquefied natural gas (LNG), Plug-in hybrid electric vehicle (PHEV), Renewable energy, Sustainable energy, Traditional centralized electricity generation.

2.1. INTRODUCTION

Early humans used natural forms of energy in their daily lives. For example, they burned wood and dry manure for cooking or heating and employed wind energy and water flow for grinding shells. In the 18th century, the use of fossil fuels resulted in the initiation of the Industrial Revolution. Humans found that fossil fuels have a higher energy density than traditional fuels. Fossil fuels and renew-

Chih-Cheng Huang & Chun-Liang Lin

able energy resources are currently used to meet the energy needs of humans. The Sun plays a role in the generation of all forms of energy, including the energy obtained from wood, wind, and fossil fuels.

2.2. Renewable and Sustainable Energy

The complex operation of the Earth is affected by the forces exerted on it by the Sun and Moon as well as geothermal energy (Fig. **2.1**). The Sun plays a role in the generation of all forms of energy on the Earth. Some of these forms of energy are as follows:

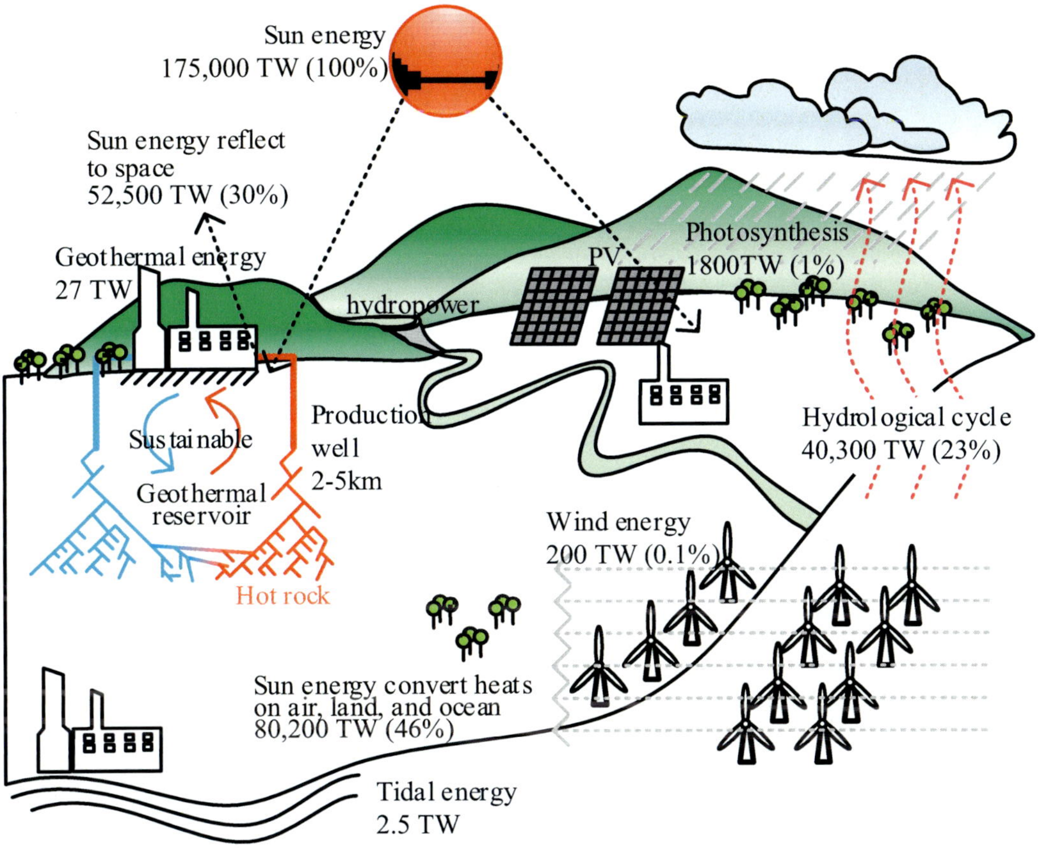

Fig. (2.1). Illustration of the formation of different forms of renewable energy.

- **Hydro Energy:** Heat radiation from the Sun results in the evaporation of seawater, which results in the hydrological cycle and rain. Electric power can be generated from water reservoirs by using turbines and generators.
- **Wind Energy:** The non-uniform irradiation of sunlight on the Earth's surface results in non-uniform temperatures on this surface. Such temperature

differences produce air and ocean currents. Air currents rotate wind turbine blades to generate electricity. In some areas, wind turbines can generate 70 kW/m of electricity [139] when the wind speed is 12 m/s.

- **Solar Energy:** Solar energy is obtained by converting light into electricity by using semiconducting materials that exhibit the photovoltaic effect [photovoltaics (PVs)]. Pakistan receives abundant sunlight with a global horizontal irradiance level of more than 1500 Wh/m2 over 90% of its area [140]. However, the efficiency of most solar panels is between 15% and 20% [141].
- **Biomass Energy:** Biomass energy is the energy (electricity or heat) produced from the plant, animal, dung, or municipal solid waste materials. The burning of plant-derived biomass results in the release of CO_2. However, biomass is nonetheless classified as a renewable energy source in the legal frameworks of the European Union and United Nations because photosynthesis results in CO_2 being cycled back into new crops.
- **Biofuel:** Biofuel refers to fuel produced from biomass through short-term processes rather than the prolonged geological processes involved in the formation of fossil fuels such as oil. If the biomass used in biofuel production can be regrown quickly, the fuel is generally considered to be a form of renewable energy.
- **Fossil Fuels:** Fossil fuels are formed through natural processes, such as the anaerobic decomposition of buried dead organisms and prehistoric algae, and they contain organic molecules that release energy on combustion. Such fuels and the dead organisms that enable their formation typically have an age of millions of years. Fossil fuel energy is more efficient than renewable energy; however, the burning of fossil fuels results in pollutant emissions and the greenhouse effect.
- **Tidal Energy:** Tides are created by the fall and rise of sea levels because of the combined effects of the Earth's rotation and the gravitational forces exerted on the Earth by the Moon and Sun. Potential energy or electricity can be generated in coastal areas from tides.
- **Geothermal Energy:** Geothermal energy refers to the thermal energy generated and stored in the Earth. The geothermal energy of the Earth's crust originates from the formation of the planet and the radioactive decay of materials. A geothermal energy plant generates electricity through steam turbines.

Various types of energy resources form the energy supply of the United States, however, not all these resources have the same environmental benefits and costs (Table **2.1**). The US voluntary market defines green power as power with environmental benefits [142].

Table 2.1. The Environmental Benefits and Costs of all Energy Resources Listed on the us Voluntary Market.

Conventional Power	Renewable Energy	
		Green Power
Oil	Large Hydropower	Wind
		Solar
Natural Gas	Municpal Solid Waste	Biomass
		Geothermal
Coal		Biogas
Nuclear		Low-Impact Hydropower
Least Benefit	More Beneficial	Most Beneficial

Conventional power refers to the power generated through the combustion of fossil fuels (*e.g*., coal, natural gas, and oil) or the nuclear fission of uranium. Fossil fuels are associated with environmental costs related to drilling, mining, and extraction. Combustion of these fuels emits greenhouse gases and causes air pollution. Although nuclear power generation does not result in the emission of greenhouse gases, it requires uranium mining and extraction and long-term radioactive waste storage.

Renewable energy refers to the energy generated from resources that are replenished over short periods. Such resources include green resources (the Sun, water, wind, biogas, biomass, and the Earth's heat) and waste materials. Some renewable energy technologies have a small impact on the environment. For example, large hydroelectric projects are associated with environmental trade-offs related to issues such as fisheries and land use.

Green power is a subset of renewable energy and is generated from renewable energy resources and technologies that provide the greatest environmental benefits. The US voluntary market defines green power as electricity produced from solar, wind, geothermal, biogas, eligible biomass, and low-impact small hydroelectric sources. Customers often buy green power for its zero-emission profile and benefits in reducing their carbon footprint.

Fig. (**2.2**) illustrates the global consumption of energy by source. This figure indicates that the share of renewables in global energy consumption has been increasing each year [143]. The global renewable energy consumption is displayed in Fig. (**2.3**). The total global consumption of renewable energy was 18 000 TWh in 2019. Renewable energy has the greatest environmental benefit, now being more environmentally beneficial than the traditional steam turbine. The global consumption of renewable energy is increasing because of strict regulations, such as universal and legally binding climate change agreements.

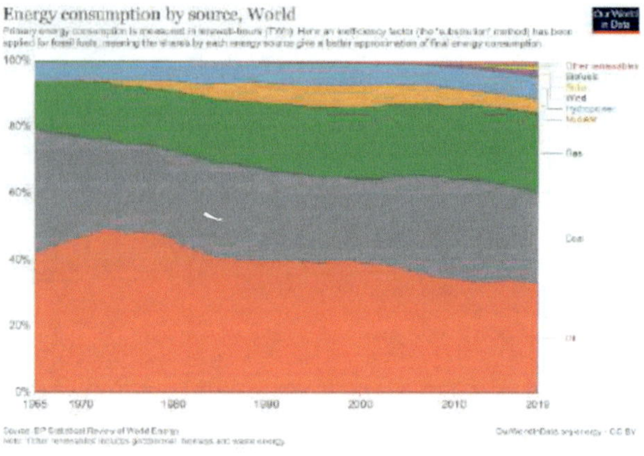

Fig. (2.2). Global energy consumption by source.

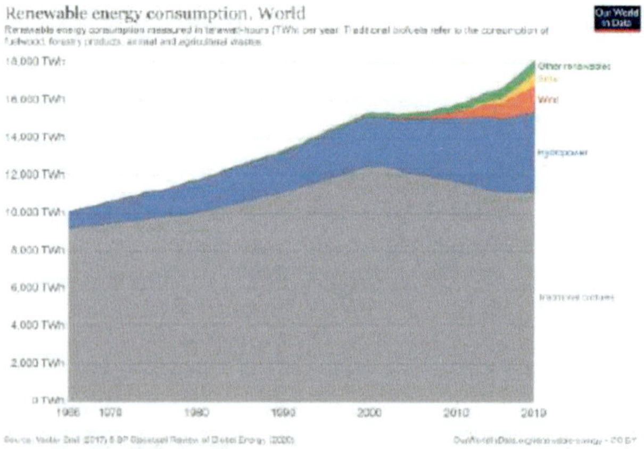

Fig. (2.3). Global renewable energy consumption.

2.2.1 Electric Energy

A schematic representation of the generation of different forms of energy is presented in Fig. (**2.4**). As depicted in Fig. (2.4 **A**), coal, liquefied natural gas (LNG), or oil is burned to release heat for rotating a steam push generator to generate electricity. However, the burning of coal, LNG, and oil results in the release of greenhouse gases; thus, fossil fuels have the least environmental benefits. As depicted in Fig. (**2.4 B**), nuclear power is generated by conducting nuclear reactions that release nuclear energy in the form of heat. Such heat is frequently used to run steam turbines for producing electricity in a nuclear power plant. Nuclear energy is a cleaner form of energy than other types of nonrenewable energy. However, nuclear reactions generate radioactive waste, which must be stored appropriately. As illustrated in Figs. (**2.4 C** and **2.4 D**), the generation of wind power and solar power has zero emissions, and these energy types thus have the greatest environmental benefits. However, such power can only be generated intermittently because of climate uncertainties.

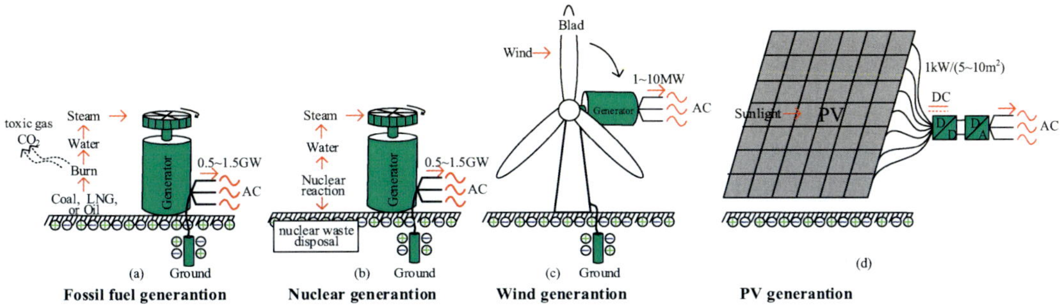

Fig. (2.4). Generation of different types of power.

2.2.2. Household Vehicle Market

Strict emission standards have diversified the development of household vehicles. In addition to high-emission internal combustion engine (ICE) vehicles, the household vehicle market offers zero-emission electric vehicles (EVs) and fuel-cell EVs as well as low-emission plug-in hybrid EVs (PHEVs). Environmental issues have attracted increasing attention; thus, many developed countries are planning to ban ICE vehicles between 2030 and 2040. Consequently, the market share of EVs has increased over time. Compared with those in 2019, global automobile sales were 23% lower in 2020 because of the COVID-19 pandemic [144]. Nevertheless, the market share of EVs continued to increase. The market share of ICE vehicles is expected to continue to decrease each year in the future, whereas that of battery EVs (BEVs) and PHEVs is expected to increase each year. The number of new BEV or PHEV purchases is expected to exceed the number of new ICE vehicle purchases from 2037 (Fig. **2.5**).

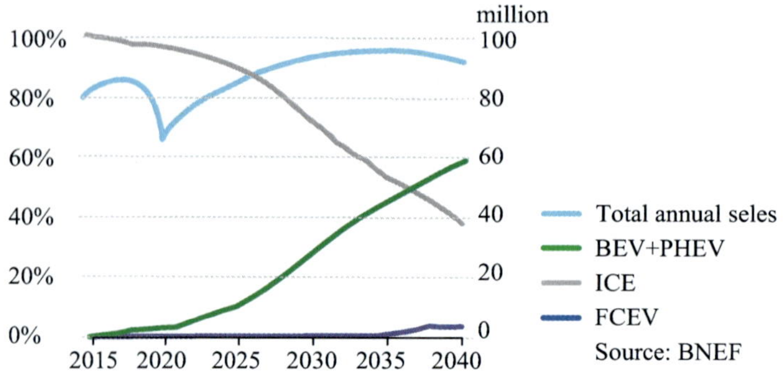

Fig. (2.5). Global shares of total annual passenger vehicle sales by drivetrain.

2.3. Optimal Unit Commitment

A power system that provides services to a section of a city comprises a central dispatch, an area dispatch, and a distribution dispatch, as shown in Fig. (**2.6**). The central dispatch contains a nuclear unit, a coal unit, and an LNG unit to generate an output voltage of 345 kV for the transmission system. The area dispatch comprises some medium-sized thermal units, three wind turbine units, and a load on a 161-kV transmission system. The distribution dispatch includes a 69-kV transmission system and 22-kV/11-kV feed system. The 69-kV transmission system contains a geothermal unit and load. The 22-kV/11-kV feed system contains two PV generators, two energy storage systems (ESSs), and loads. The aforementioned three hierarchical dispatches are linked by transformers.

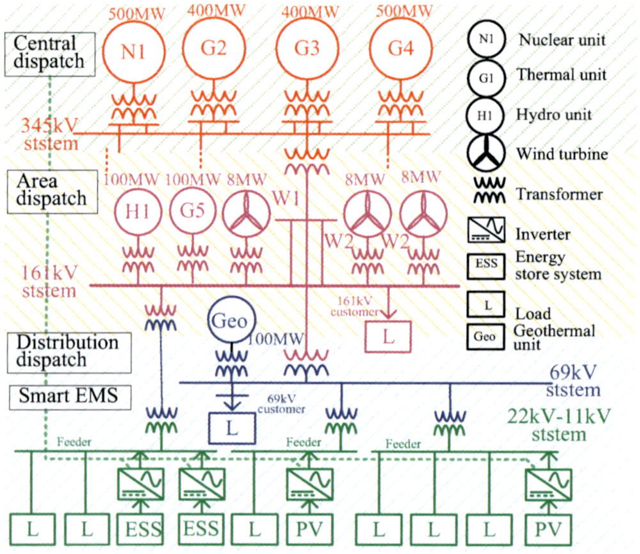

Fig. (2.6). Structure of a power system that provides services to a section of a city.

The central large thermal units, such as based units, enable high output power generation at a low cost; however, these units cause pollution and have a long response time, including long ramp-up and ramp-down times. Medium-sized thermal units, such as the oil unit, enable rapid response when inrush load is present; however, they cause pollution and are associated with high power generation costs. The power system described in the aforementioned text can also include green power units, such as wind turbine units, PV generators, geothermal units, and hydro units, which enable the production of zero-emission energy.

The aforementioned power system has a power rate of 2.1 GW and can support the power needs of 10 000 people in a city. All the units in the aforementioned system do not have the same cost and transmission loss. Therefore, a suitable commitment must be made for all units to ensure efficient power generation. Historical records, statistics, and optimization algorithms should be used to select a suitable commitment to the spinning reserves for the central large and medium-sized thermal units for achieving optimal economic electricity generation to meet load requirements [145 – 147]. In each unit, an automatic generation control system is employed to meet the unit's commitment requirements [148 – 150] with a stable voltage and frequency. However, green power is generated intermittently; thus, generating green power economically becomes complicated, and the power system must contain additional spinning reserves to compensate for the intermittent generation of green power [151 – 153]. In general, the spinning reserve units are oil units or quick response units.

Fortunately, with the progress of battery technology, reducing battery and ESS costs has become possible. Affordable electricity storage devices would enhance the efficiency of green power generation systems and the adoption of green power. The global average battery capacity of each EV is 50 kWh [154]. If a city has 1000 EVs, 50 MWh of energy would be required to run these vehicles. For this energy, a spinning reserve of 10% over 1h would be sufficient. An aggregator administrates numerous ESS devices to make this power system parallel; thus, this power system is a smart electricity management system. The unit commitment of the ESS helps an independent system operator implement optimal electricity dispatch [155, 156]. Green power is associated with lower emissions than power generated from fossil fuels.

CONCLUSION

Green power, such as wind and solar power, provides maximum environmental benefits and is associated with zero emissions and low pollution. However, because of the large construction area and costs required for generating green power, it accounts for only less than 3% of global power generation.

The household vehicle market offers two main types of vehicles, ICE vehicles and EVs. However, many developed countries are planning to ban ICE vehicles between 2030 and 2040. EVs are efficient and have zero emissions. However, their use is limited by their high battery costs and long charging time. Moreover, the batteries of EVs must be charged with considerable electrical energy. The cost strategy involves charging the battery during the off-peak hours. A large ESS distributes the commitment in an optimal manner over the various system units and can thus increase the quantity of green power in a power grid.

Reducing battery costs is critical for increasing the share of green power in the power grid and the use of EVs in the future.

CHAPTER 3

Electromagnetic Waves

Abstract: Mobile communication, which is a crucial aspect of daily life, is based on electromagnetic waves, which transmit not only information but also energy. Electromagnetic waves can be transmitted wirelessly to facilitate wireless power transfer for charging vehicle batteries. Wireless charging with electromagnetic waves does not require the use of a plug point for convenience and safety. The British scientist Maxwell discovered that electromagnetic waves can be radiated, and such radiation is a natural phenomenon caused by the interaction between electricity and magnetism. Moreover, the British Scientist Faraday and French physicist Ampere found that magnetism generates electricity and that electricity generates magnetism, respectively. This chapter discusses electromagnetic wave radiation on the basis of Faraday's theorem, Ampere's theorem, and Maxwell's equation.

Keywords: Ampere's law, Electromagnetic wave radiation, Faraday's law, Maxwell's equations, Mobile communication, Wireless charging.

3.1. INTRODUCTION

The electromagnetic spectrum is displayed in Fig. (**3.1**). The frequency band for wireless power charging is between 10 kHz and 20 MHz. These frequencies are in the radio waveband and are associated with ultralow quantum energy. Therefore, wireless power charging does not adversely affect the human body [157 – 159].

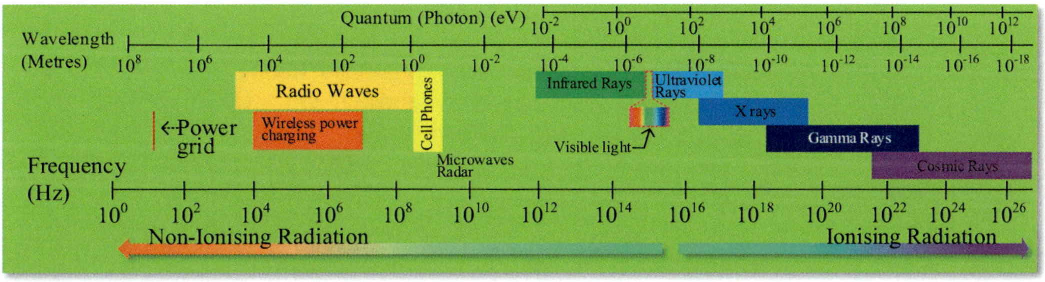

Fig. (3.1). Electromagnetic spectrum.

Chih-Cheng Huang & Chun-Liang Lin

In wireless power charging, an inverter is used to generate high-frequency output current and a synchro electromagnetic field to transfer power through an inductor–capacitor tank (LC tank) [160 – 162], which contains capacitors and inductors. Different quality factor (*Q* factor) designs are used in different applications [163, 164].

For electromagnetic induction power transfer applications (performed at less than 100 kHz), an LC tank with a low *Q* value and a power transfer distance of less than 1 m is designed to achieve high electrical energy transmission for mass production. For novel magnetic resonance wireless power transfer applications (performed at higher than 100 kHz), an LC tank with a transmission distance of several meters and a relatively high *Q* factor is usually designed.

LC tanks with different transmission distances, operation frequencies, and designs are used in wireless power transfer applications [165 – 167].

Faraday's law indicates that a time-varying magnetic field produces a voltage across an active coil's two endpoints. A typical application of this law is a generator. In a generator, kinetic energy is generated by rotating magnetic fields to intersect an active coil through the air gap. This kinetic energy causes the rotation of a rotor containing magnets, which results in the generation of a back electromotive force (EMF) that acts on the two endpoints of the active coil. Electromagnetic induction has uses in daily life, such as in induction cookers, which use magnetic fields to generate eddy currents for heating.

Ampere's law states that a current can produce a magnetic field. Typical applications of Ampere's law include electromagnets, relays, and door latches. Maxwell revised Ampere's law and proved that a time-varying electric field radiates outward and generates alternating electric and magnetic fields. Wireless power transfer is based on Maxwell's equations, which are described in the following section.

3.2. Maxwell's Equations

A young Scottish scientist named Maxwell published an article that summarized the law of electromagnetics and extended Ampere's law to make it more complete. This article became a classical work in electromagnetics and made Maxwell a top researcher in the field of electromagnetism.

Scientists simplified the research presented in the aforementioned article into four classical equations, which are called Maxwell's equations. These four equations accurately describe electromagnetic phenomena, and their importance is comparable to that of Einstein's theory of relativity and Newton's law of universal gravitation.

Maxwell's equations are the core of electromagnetics because all electromagnetic phenomena can be expressed using electricity and magnetism vectors [168 – 170]. Maxwell's equations describe electromagnetic phenomena, such as the generation of magnetism by electricity, the generation of electricity by magnetism, the electric charge being the source of electric flux, and the absence of a single positive or negative magnetic charge in nature. This chapter explains the physical meaning of Maxwell's equations.

3.2.1. Faraday's law

Maxwell's first equation is called Faraday's law and the law of electromagnetic induction. This law is expressed as follows:

$$\nabla \times \vec{E} = -\frac{\partial \vec{B}}{\partial t} \tag{3.1}$$

where $\vec{E}$ is the electric field intensity, $\nabla \times \vec{E}$ is the curl [171 – 173] of $\vec{E}$, and $\frac{\partial \vec{B}}{\partial t}$ is the partial differential of the magnetic flux density $\vec{B}$. A visual representation of equation 3-1 is displayed in Fig. (**3.2**), in which vector field $\vec{E}$ at any point P is generated from $-\frac{\partial \vec{B}}{\partial t}$. The direction of rotation of vector field $\vec{E}$ is determined using the right-hand rule. When the time-varying magnetic flux density $\left|-\frac{\partial \vec{B}}{\partial t}\right|$ increases, the electric field intensity $\vec{E}$ also increases. Thus, the electric field intensity is produced by the time-varying magnetic flux density $\bar{B}$. This electromagnetic effect is illustrated in Fig. (**3.3**).

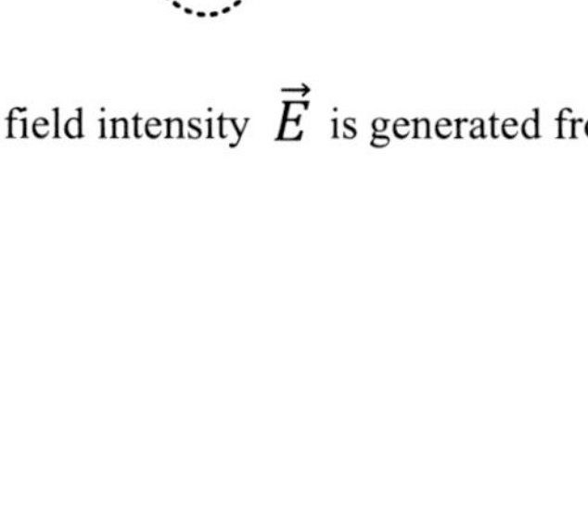

Fig. (3.2). Faraday's law (the electric field intensity $\vec{E}$ is generated from the magnetic flux density $-\frac{\partial \vec{B}}{\partial t}$).

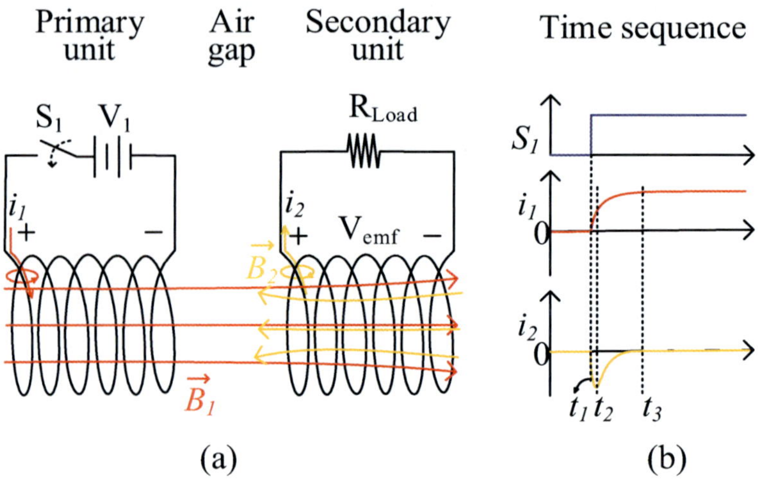

Fig. (3.3). Faraday's electromagnetic effect.

When coil S_1 of the primary unit Fig. (**3.3a**) is turned on, currents i_1 and i_2 are generated at time t_1, as shown in the time sequence in Fig. (**3.3b**). Currents i_1 and i_2 synchronously generate magnetic flux densities $\overrightarrow{B_1}$ and $\overrightarrow{B_2}$, the directions of which adhere to the right-hand law. The direction of the vector field $\overrightarrow{B_2}$ is opposite to that of $\overrightarrow{B_1}$, and $\overrightarrow{B_2}$ resists the changing of $\overrightarrow{B_1}$ as per Lenz's law [174, 175]. The maximum time-varying rate of current i_1 and the peak value of current i_2 occur at time t_2. Next, the time-varying rate of current i_1 gradually decreases. In this case, current i_2 decays between t_2 and t_3. Between t_1 and t_3, an electric field is generated from the time-varying magnetic flux, as per Faraday's law. The EMF voltage (V_{emf}) is expressed as follows:

$$V_{emf} = -\frac{d\Phi B_1}{\mathrm{dt}} \tag{3.2}$$

where V_{emf} is generated from the time-varying $-\Phi B_1$ [176, 177].

Power companies use kinetic energy to drive generator rotors, which produce a magnetic field that intersects a coil on a stator to generate a voltage at the coil's endpoint. The application of Faraday's law in a generator is illustrated in Fig. (**3.4**). Power companies use power transmission and distribution systems, including towers and transformers, to convert voltages into an AC voltage of 100 or 200 V.

From the physical perspective, Faraday's law reveals that electrical energy can be generated from a magnetic field as long as a time-varying magnetic flux intersects the coil.

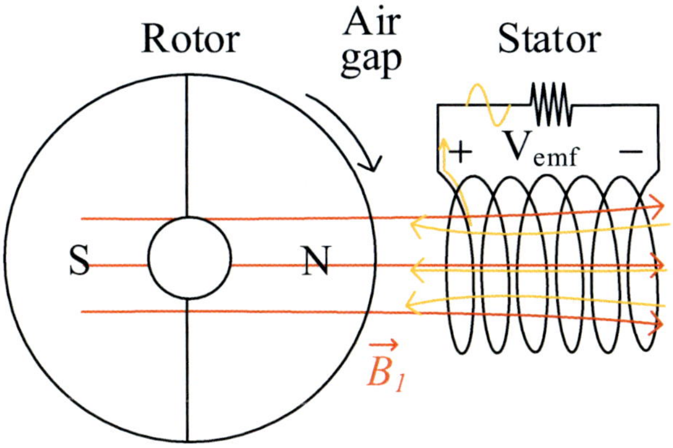

Fig. (3.4). Electromagnetic application of Faraday's law in a generator.

3.2.2. Ampere's law

Maxwell's second equation was discovered by the French physicist Andre-Marie Ampere; therefore, this equation is called Ampere's law, which is expressed as follows:

$$\nabla \times \vec{H} = \vec{J} + \frac{\partial \vec{D}}{\partial t} \quad \textbf{(3.3)}$$

where $\vec{H}$ is the magnetic field intensity, $\nabla \times \vec{H}$ is the curl of $\vec{H}$, $\vec{J}$ is the current density, and $\vec{D}$ is the partial differential of the electric flux density $\vec{D}$ with respect to the time t. Equation 3-3 indicates that the magnetic field intensity $\vec{H}$ is generated from the current in the conductor and the time-varying electric flux density. The current I in the conductor generates magnetic field intensity $\vec{H}$, as illustrated in Fig. (**3.5a**). In this figure, the $\vec{H}$ plane is perpendicular to the current I, and the right-hand rule determines the rotation direction of $\vec{H}$. If the current increases, the magnetic field intensity $\vec{H}$ also increases. Because a conductor contains many free electrons, its current density is considerably higher than $\frac{\partial \bar{D}}{\partial t}$ therefore, equation 3-3 is simplified as follows:

$$\nabla \times \vec{H} = \vec{J} \quad \textbf{(3.4)}$$

The simple circuit displayed in Fig. (**3.5b**) indicates that the conductor's current generates a magnetic field. In the aforementioned figure, *E* is the energy, S_1 is the current switch and *R* is the resistance, which changes *I*. The time sequence diagram corresponding to the aforementioned circuit is depicted in Fig. (**3.5c**). When S_1 is turned on at t_1, the DC voltage generates constant current *I* flowing through the resistor, and magnetic field intensity $\vec{H}$ is generated in the wire. When S_1 is turned off at t_2, the current field intensity and magnetic field intensity are 0. Equation 3-4 indicates that the magnetic field intensity $\vec{H}$ originates from the current density $\vec{J}$.

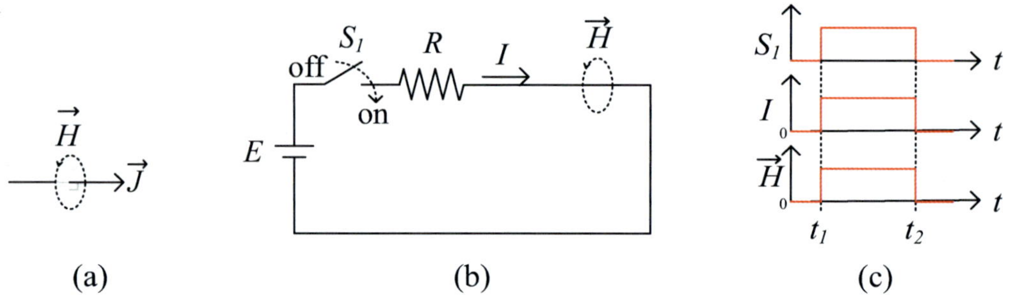

Fig. (3.5). Generation of magnetic field intensity $\vec{H}$ from constant current *I*.

The time-varying electric flux density $\vec{D}$ generates magnetic field intensity $\vec{H}$, as displayed in Fig. (**3.6a**). In this figure, the $\vec{H}$ plane is perpendicular to the electric flux density $\vec{D}$. When $\vec{D}$ increases, $\vec{H}$ also increases. This phenomenon occurs in insulators. Because almost no free electrons exist in an insulator, the current density approaches 0 ($\vec{J} \approx 0$); thus, equation 3-3 can be simplified as follows:

$$\nabla \times \vec{H} = \frac{\partial \vec{D}}{\partial t} \tag{3.5}$$

The simple circuit depicted in Fig. (**3.6b**) indicates that the electric flux density generates magnetic field intensity. In this figure, *v*(*t*) is the energy, S_1 is the electric flux density switch and *R* is the resistance, which changes $\vec{D}$. The two parallel metal sheets in Fig. (**3.6b**) charge or discharge similar to that of a capacitor. The time sequence diagram for the aforementioned circuit is displayed in Fig. (**3.6c**). When S_1 is turned on at t_1, the source of *v*(*t*) generates energy and transfers it to the two parallel metal sheets through the resistor, which results in electric flux density $\vec{D}$ between the two parallel metal sheets. In this scenario, the $\vec{H}$ plane exists and is perpendicular to $\vec{D}$. When S_1 is turned off at t_2, the electric flux density and magnetic field intensity are 0.

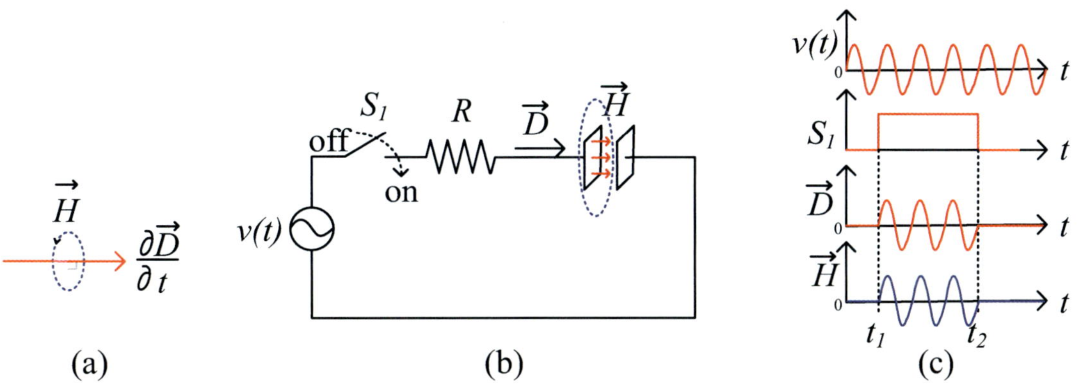

Fig. (3.6). Generation of magnetic field intensity $\vec{H}$ from time-varying electric flux density $\vec{D}$.

On the basis of the aforementioned description, Ampere's law indicates that a magnetic field can be generated using the following two methods:

1. Magnetic field intensity $\vec{H}$ is generated from current I; that is, $\nabla \times \vec{H} = \vec{J}$.
2. Magnetic field intensity $\vec{H}$ is generated from the time-varying electric flux density $\vec{D}$; thus, $\nabla \times \vec{H} = \frac{\partial \vec{D}}{\partial t}$.

Equation 3.5 is a corrected equation presented by Maxwell's second equation and is the theoretical basis for the outward radiation of electromagnetic waves.

3.2.3. Gauss' law

Maxwell's third equation, which is also called Gauss' law, is expressed as follows:

$$\nabla \cdot \vec{D} = \rho \qquad \textbf{(3.6)}$$

where $\nabla \cdot \vec{D}$ is the divergence of the electric flux density $\vec{D}$ and ρ is the charge density. Equation 3-6 indicates that electric flux density is generated by an electric charge. When $\rho > 0$, $\vec{D}$ diverges outward, and when $\rho < 0$, $\vec{D}$ converges inward. The electric flux converges inward for a negative charge, which results in nearby free electrons being repelled from this charge (Fig. **3.7a**). Conversely, the electric flux diverges outward for a positive charge, which results in nearby free electrons being attracted to this charge (Fig. **3.7a**).

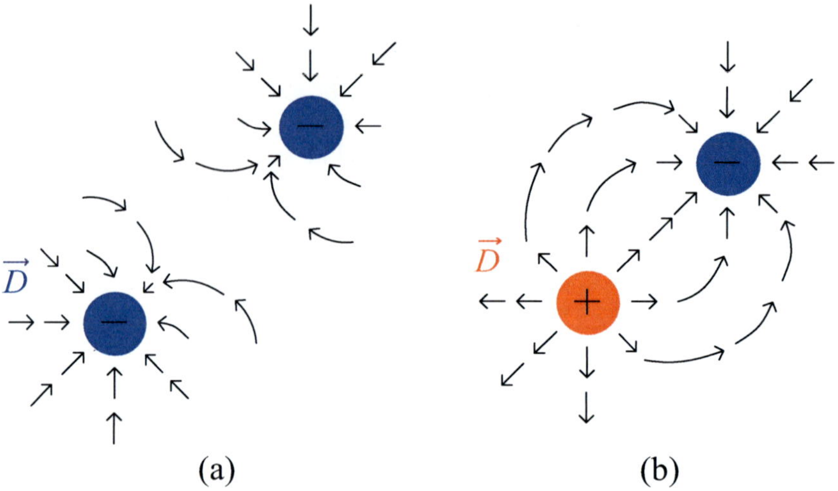

Fig. (3.7). Generation of electric flux density from charges.

3.2.4. Gauss' Law for Magnetism

Maxwell's fourth equation, which is also called Gauss' law for magnetism, is expressed as follows:

$$\nabla \cdot \vec{B} = 0 \tag{3.7}$$

The right-hand side of equation 3-7 is 0, which indicates that the divergence of the magnetic flux density $\vec{B}$ is always equal to 0. These studies have found that no magnetic monopoles exist in nature. All existing magnetic poles are dipoles; therefore, the divergence of $\vec{B}$ is 0. Equation 3-7 indicates that the outward magnetic flux density is equal to the inward magnetic flux at any point (Fig. **3.8**). Fig. (**3.8**) shows that the entering magnetic flux density $\overrightarrow{B_S}$ is equal to the exiting magnetic flux density $\overrightarrow{B_N}$ at any point in space; therefore, the divergence of $\vec{B}$ is always 0. The fundamental difference between $\vec{B}$ and $\vec{D}$ lies in the difference in their divergences.

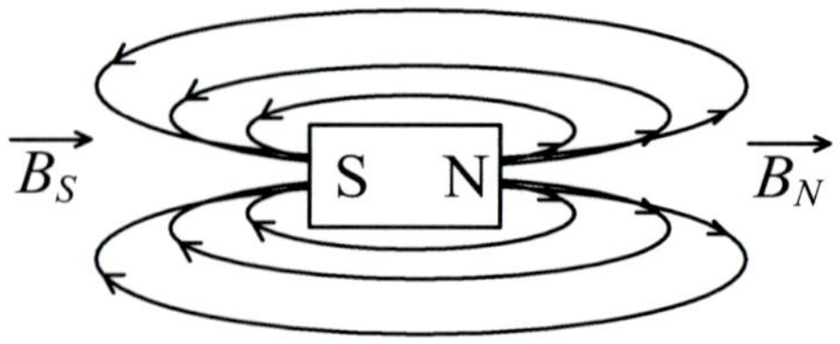

Fig. (3.8). Magnetic dipole.

3.3. Wave Equation

Ampere discovered that a current can generate a magnetic field, and the higher the current, the stronger the generated magnetic field. Maxwell made considerable amendments to Ampere's law and proved that a time-varying electric field also generates a magnetic field. This discovery indicates that time-varying electric and magnetic fields are derived from each other and form electromagnetic waves that radiate outward in the air.

3.3.1. Concept and principle of the Wave Equation

The relationship between the electric field intensity $\vec{E}$ and electric flux density $\vec{D}$ is as follows:

$$\vec{D} = \epsilon\vec{E} \tag{3.8}$$

where ϵ is the electromagnetic field's permittivity in the medium and μ is the permeability of the medium. The relationship between the magnetic field intensity $\vec{H}$ and magnetic flux density $\vec{B}$ is as follows:

$$\vec{B} = \mu\vec{H} \tag{3.9}$$

Therefore, if $\vec{E}$ and $\vec{H}$ are known, $\vec{D}$ and $\vec{B}$ can be easily determined using equations 3-8 and 3-9, respectively. Consequently, we focus on finding $\vec{E}$ and $\vec{H}$ and analyzing their effect by using Maxwell's equation. Because $\vec{B} = \mu\vec{H}$, equation 3-1 can be revised as follows:

$$\nabla \times \vec{E} = -\mu\frac{\partial\vec{H}}{\partial t} \tag{3.10}$$

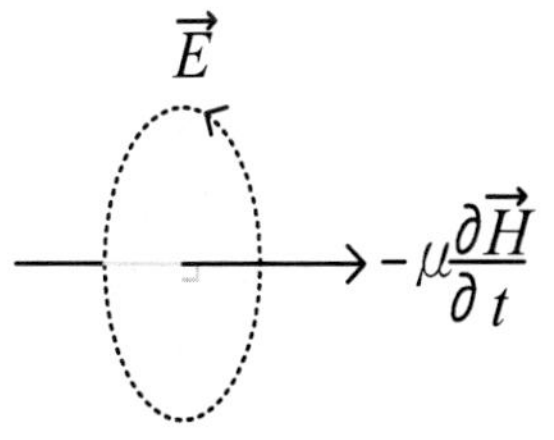

Fig. (3.9). Generation of the curl of electric field intensity $\vec{E}$ by $-\mu\frac{\partial\vec{H}}{\partial t}$.

In Fig. (**3.9**), the time-varying magnetic field intensity $\vec{H}$ is directed along the *x*-axis. In the plane perpendicular to the *x*-axis, $\vec{H}$ induces electric field intensity $\vec{E}$, which is proportional to $\frac{\partial \vec{H}}{\partial t}$ this indicates that a magnetic field with more rapid intensity variation results in higher electric field intensity $\vec{E}$.

Equation 3-10 indicates that $\vec{H}$ induces electric field intensity $\vec{E}$, which is the essential characteristic of the electromagnetic field. Because $\vec{D} = \epsilon \vec{E}$, equation 3-3 can be revised as follows:

$$\nabla \times \vec{H} = \vec{J} + \epsilon \frac{\partial \vec{E}}{\partial t} \quad \textbf{(3.11)}$$

where $\vec{J}$ is the current density. The second term on the right-hand side of equation 3-11 is the correction term added by Maxwell for the radiation of electromagnetic waves. For a conductor, equation 3-11 can be rewritten in a relatively simple form. Because the conductivity of a conductor is high, the current density $\vec{J}$ in the conductor is considerably larger than $\epsilon \frac{\partial \vec{E}}{\partial t}$ thus, equation 3-11 can be simplified as follows:

$$\nabla \times \vec{H} = \vec{J} \quad \textbf{(3.12)}$$

In Fig. (**3.10**), the current is directed along the *x*-axis. In the plane perpendicular to the *x*-axis, the current induces magnetic field intensity $\vec{H}$. The magnetic field intensity $\vec{H}$ is proportional to the current density $\vec{J}$. Equation 3-12 indicates that the current induces magnetic field intensity $\vec{E}$.

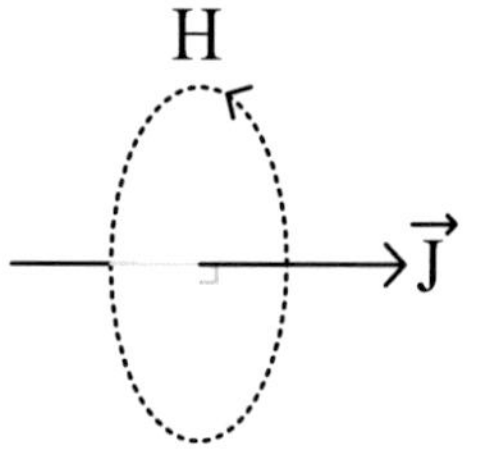

Fig. (3.10). Generation of magnetic field intensity $\vec{H}$ by current $\vec{J}$.

Because the conductivity of an insulator is very low, $\vec{J}\approx 0$; thus, equation 3-11 can be simplified as follows:

$$\nabla \times \vec{H} = \epsilon \frac{\partial \vec{E}}{\partial t} \qquad \textbf{(3.13)}$$

In Fig. (**3.11**), the time-varying electric field intensity $\vec{E}$ is directed along the *x*-axis. In the plane perpendicular to the *x*-axis, $\vec{E}$ induces magnetic field intensity $\vec{H}$. The magnetic field intensity $\vec{H}$ is proportional to $\frac{\partial \vec{E}}{\partial t}$, which indicates that an electric field with more rapid intensity variation generates higher magnetic field intensity $\vec{H}$. This finding is similar to that obtained for the generation of a current by a magnetic field.

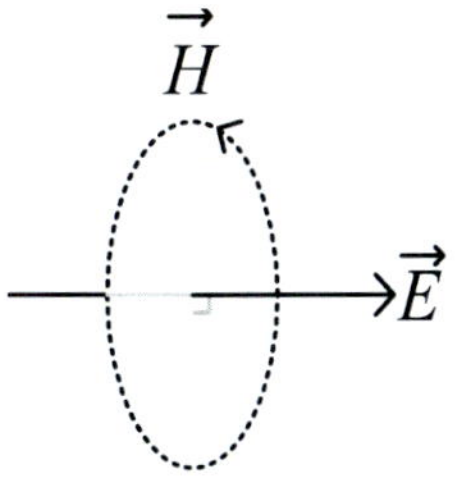

Fig. (3.11). Generation of magnetic field intensity $\vec{H}$ by time-varying electric field intensity $\vec{E}$.

Figs. (**3.9** and **3.11**) clearly illustrate the relationship between $\vec{H}$ and $\vec{E}$. Time-varying $\vec{H}$ induces electric field intensity $\vec{E}$, and time-varying $\vec{E}$ induces magnetic field intensity $\vec{H}$. The mutual induction of an electric field and a magnetic field is the essential characteristic of an electromagnetic field.

The magnetic field $\vec{H}$ and electric field $\vec{E}$ radiate outward in space, with the angle between $\vec{H}$ and $\vec{E}$ being 90°, as shown in Fig. (**3.12**).

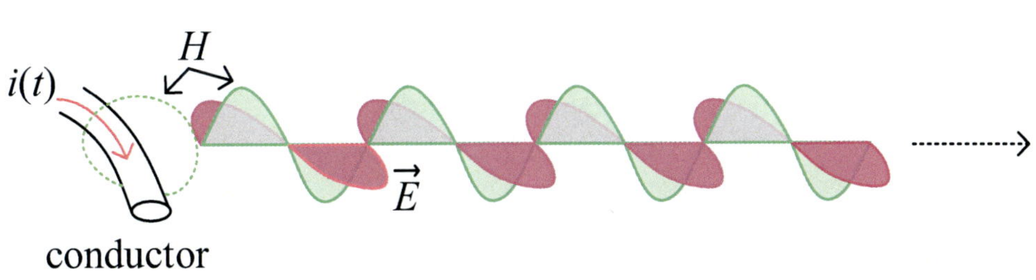

Fig. (3.12). Outward radiation of the magnetic field and electric field in space.

An electric field and a magnetic field interact as per Faraday's and Ampere's laws to form an electromagnetic wave that radiates outward. The existence of electromagnetic waves was speculated by Maxwell and proven by the German scientist H. R. Hertz 20 years later. Electromagnetic waves are the basis of modern wireless communication.

CONCLUSION

This chapter describes electromagnetic waves by using Maxwell's four equations, namely Faraday's law, Ampere's law, and Gauss's two laws.

Faraday's law is illustrated in Fig. (**3.3**). In this figure, the source and load are not connected by a wire. After the switch S_1 is turned on, a time-varying magnetic field $\overrightarrow{B_1}$ is generated on coil L_1 on the source side. Coil L_2 on the load side simultaneously generates a time-varying magnetic field $-\overrightarrow{B_1}$, which acts in the opposite direction to $\overrightarrow{B_1}$. The time-varying magnetic field $-\overrightarrow{B_1}$ generates a back EMF (V_{emf}) on L2, as per equation (3-1), which enables electrical energy to be transferred from the source to the load between t_1 and t_3. If electrical energy is transmitted continuously, the source continuously generates $\overrightarrow{B_1}$. Thus, the switch S_1 on the source side can be replaced by an AC source to generate $\overrightarrow{B_1}$ continuously.

Ampere's law is illustrated in Figs. (**3.5** and **3.6**). In these figures, the time-varying current generates magnetic field intensity $\vec{H}$. Moreover, magnetic field intensity $\vec{H}$ generates time-varying current *i*(t), as illustrated in Fig. (**3.12**). As displayed in Fig. (**3.12**), the magnetic field $\vec{H}$ and electric field $\vec{E}$ interact continuously to produce electromagnetic waves continuously.

Therefore, wireless power transfer is based on Faraday's law and Ampere's law.

CHAPTER 4

AC Network Analysis

Abstract: Wireless power transfer for vehicle charging is inducted between two power pads. A power pad is combined with a tuned network (compensation network) to construct a resonant circuit for efficiently transmitting and receiving large quantities of energy. State-of-the-art wireless power transfer through inductive and magnetic resonances has numerous potential applications. A simple tuned network with a series or parallel inductor–capacitor (LC) circuit generates appropriate quality factors (*Q* factors) for various applications, such as induction cookers. Because of misalignment and a varying air gap between two power pads, specially tuned networks, such as inductor–capacitor–inductor and inductor–capacitor–capacitor networks, have been developed for various applications, such as electric vehicle charging. A tuned network with a high *Q* factor can achieve long-distance wireless energy transmission. A prototype system with magnetic resonance has been developed. This network can wirelessly transmit power of 60 W with an efficiency of approximately 40% over a distance of more than 2 m. A tuned network is constructed from an AC network, which contains inductors and capacitors. Different types of LC networks produce different effects. A series resistor–inductor–capacitor (RLC) resonant circuit can amplify voltage, whereas a parallel RLC resonant circuit can amplify current. This chapter describes the resonance principle, *Q* factor design, bandwidth, and selectivity of series and parallel RLC circuits.

Keywords: Inductive resonance, Inductor–capacitor–capacitor (LCC) network, Inductor–capacitor–inductor (LCL) network, Magnetic resonance, Quality factor (*Q* factor), Resonant circuit, Tuned network.

4.1. INTRODUCTION

The wireless charging of electric vehicles (EVs) is achieved through inductive power transfer [178 – 181]. The circuit displayed in Fig. (**4.1**) contains the following two units: a primary unit and a secondary unit. Power flows from the primary unit to the secondary unit through two tuned networks. The tuned network of the primary unit contains a capacitor C_p and an inductor L_p that are connected in parallel. The tuned network of the secondary unit contains a capacitor C_s and an inductor L_s that are connected in series. In the primary unit, low-frequency voltage from the grid is converted into DC voltage by a rectifier and filter. This DC voltage acts as the power source of the inverter in the primary unit. The inverter outputs high-frequency current into the tuned network of the

Chih-Cheng Huang & Chun-Liang Lin

primary unit to link this unit with the secondary unit's tuned network through mutual inductance. The secondary unit receives power from the primary unit, and this power is transferred to the load through a rectifier and filter.

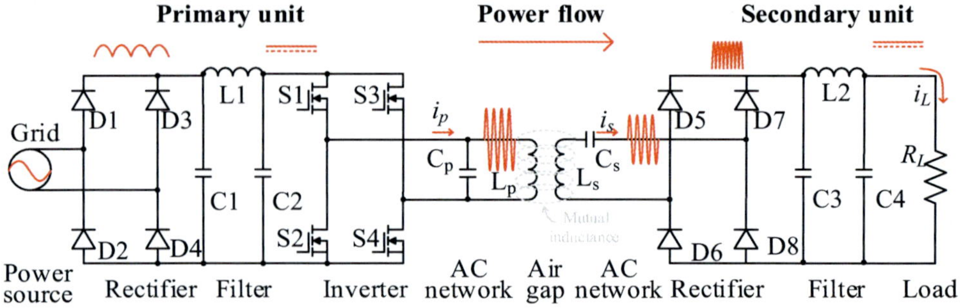

Fig. (4.1). Wireless power transfer in an EV's battery charger.

A resonating tuned network implements minimum Voltage Ampere (VA) [102, 112] so that it can transfer considerable energy with high efficiency. Because of misalignment and a varying air gap between two power pads, complex tuned networks, such as inductor–capacitor–inductor (LCL) [109, 182, 183] and induc tor– capacitor–capacitor (LCC) networks, have been developed for various applications. A tuned network with a simple structure and a suitable quality factor (*Q* factor) can be used in various applications, such as an inductive heater [186]. A tuned network with a high *Q* factor and magnetic resonance technology can achieve long-distance wireless power transfer [187].

Tuned networks perform critical tasks for efficiency on long-distance power transmission. A tuned network is constructed from an AC network, which contains inductors and capacitors [188, 189]. Different types of inductor-capacitor (LC) networks have different effects. For example, a series resistor-inductor–capacitor (RLC) resonant circuit amplifies voltage, whereas a parallel RLC resonant circuit amplifies current.

This chapter describes the fundamental components of an RLC circuit, namely resistors, inductors, and capacitors, as well as the resonance principle, *Q* factor design, bandwidth (BW), and selectivity of series and parallel RLC circuits.

4.2. Fundamental Components and Power Consumption of Resonant Circuits

The fundamental components of electronic circuits include resistors, capacitors, and inductors. Resistors consume power, whereas capacitors and inductors store energy. The resistor's resistance is a constant that does not change with the voltage frequency. However, the capacitor's capacitive reactance and the inductor's inductive reactance vary with the voltage frequency. A simple electronic circuit is displayed in Fig. **4.2 (a)**. In this figure, an AC power source supplies load (*Z)*. In Fig. **4.2 (b)**, Cartesian coordinates are used to represent the relationship among impedance (*Z*), resistance (*R*), and reactance (*X*). The impedance *Z* comprises a real part and an imaginary part. The real part of the impedance is the resistance *R*, whereas its imaginary part is the inductive reactance X_L. The impedance *Z* is expressed in equation 4.1, and $||Z||$ is expressed in equation 4.2.

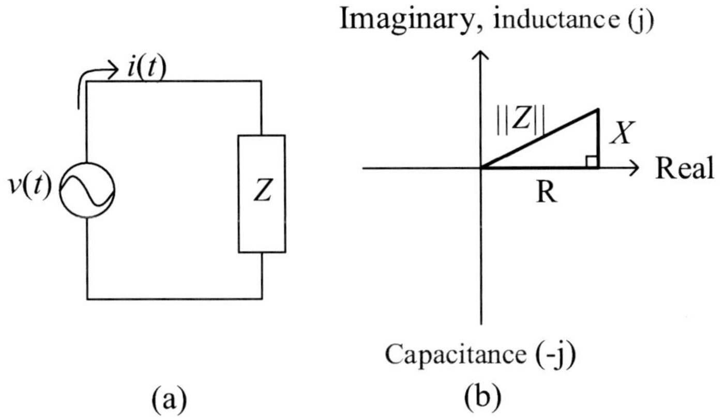

Fig. (4.2). Simple electronic circuit with impedance (*Z*) represented using Cartesian coordinates.

$$Z = R + j\,X_L \qquad \textbf{(4.1)}$$

$$||Z|| = \sqrt{R^2 + X_L^2} \qquad \textbf{(4.2)}$$

4.2.1. Resistor

A resistor is a conductor that hinders current flow by consuming power. A resistor and its symbol are shown in Fig. (**4.3**), in which v_R and i_R represent the voltage across and the current flowing through the resistor, respectively. A resistor's resistance is expressed in equation 4.3.

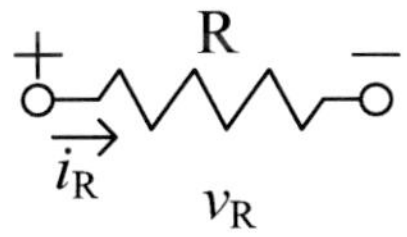

Fig. (4.3). Resistor and its electric symbol.

$$R = \frac{v_R}{i_R} \qquad \textbf{(4.3)}$$

4.2.2. Inductor

Inductors generally comprise copper wires arranged in a spiral shape that can store electrical energy. A greater number of turns of the spiral corresponds to higher inductance. An inductor and its symbol are depicted in Fig. (**4.4**), in which v_L and i_L represent the voltage across and current flowing through the inductor, respectively. The inductive voltage of an inductor (v_L) is expressed in equation 4.4. The derivation of the inductive reactance X_L is presented in equations 4.5 –4.8. Equation 4.8 indicates that when the frequency of v_L (ω) increases, the inductive reactance also increases.

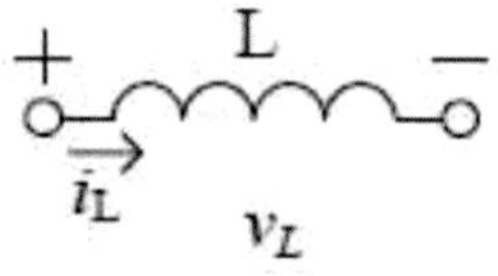

Fig. (4.4). Inductor and its electric symbol.

$$v_L(t) = L\frac{di}{dt} \tag{4.4}$$

$$v_L(t) = L\frac{d}{dt}i_{peak}\, sin\, \omega t \tag{4.5}$$

$$v_L(t) = \omega L\, i_{peak}\, cos\, \omega t \tag{4.6}$$

$$V_{peak} = \omega L\, i_{peak} \tag{4.7}$$

$$X_L = \frac{v_{peak}}{i_{peak}} = \omega L \tag{4.8}$$

4.2.3. Capacitor

A capacitor comprises two metal sheets that have a small air gap between them, which can store electrical energy. A greater metal sheet area or smaller air gap between the metal sheets corresponds to a greater electric capacity. A capacitor and its symbol are illustrated in Fig. (**4.5**), in which v_C and i_C represent the voltage across and the current flowing through the capacitor, respectively. The current i_C is expressed in equation 4.9. The derivation of the capacitive reactance X_C is presented in equations 4.10 –4-13. Equation 4.13 indicates that a higher frequency of i_C (ω) corresponds to lower capacitive reactance.

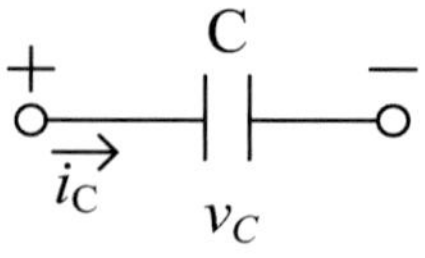

Fig. (4.5). Capacitor and electric symbol.

$$i_C(t) = C\frac{dv_C}{dt} \quad (4.9)$$

$$i_C(t) = C\frac{d}{dt}\, v_{peak} \sin \omega t \quad (4.10)$$

$$i_C(t) = \omega C\, v_{peak} \cos \omega t \quad (4.11)$$

$$i_{peak} = \omega C\, v_{peak} \quad (4.12)$$

$$X_C = \frac{v_{peak}}{i_{peak}} = \frac{1}{\omega C} \quad (4.13)$$

4.3. Resonant Circuit

A resonant circuit can amplify voltage or current, which causes high power and high-efficiency in wireless power transfer.

4.3.1. Series RLC resonant circuit

A series RLC circuit is displayed in Fig. (**4.6**), and the impedance Z of this circuit is expressed in equation 4.14. When the source voltage's frequency is f_0, resonance occurs in the circuit. This phenomenon results in a minimum Z value, as presented in equation 4.15. The resonant frequency f_0 is expressed in equations 4.16) and 4.17). The relationships of X_L, X_C, and Z with the source voltage's frequency are illustrated in Fig. (**4.7**), in which X_{L0} and X_{C0} are the inductive reactance and capacitive reactance at f_0, respectively.

$$\begin{aligned} Z &= R + jX_L - jX_C \\ &= R + j(X_L - X_C) \end{aligned} \quad (4.14)$$

$$\min. Z = R + j\left(2\pi f_0 L - \frac{1}{2\pi f_0 C}\right) = R \quad (4.15)$$

$$2\pi f_0 L = \frac{1}{2\pi f_0 C} \quad (4.16)$$

$$f_0 = \frac{1}{2\pi\sqrt{LC}} \quad (4.17)$$

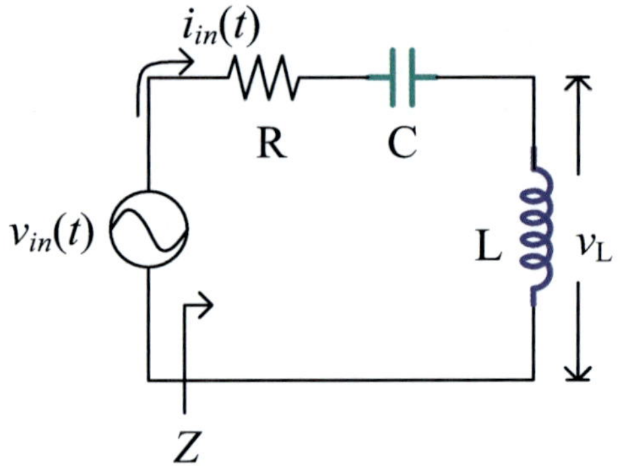

Fig. (4.6). RLC series circuit.

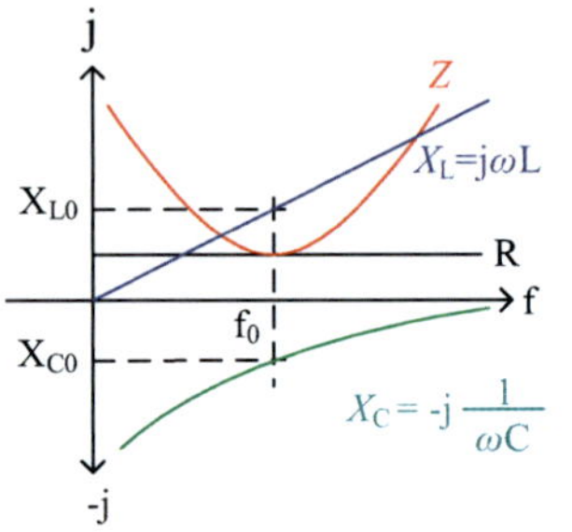

Fig. (4.7). Relationships of X_L, X_C, and Z with the source voltage's frequency.

4.3.2. Power

The *I–f* curve of a series RLC circuit is displayed in Fig. (**4.8**). In this figure, the imaginary part of the impedance Z represents the capacitive reactance when $f < f_0$ because $X_C > X_L$ under this condition. The imaginary part of Z is 0, when $f = f_0$ because $X_C = X_L$ under this condition. The maximum current also occurs when $f = f_0$. The imaginary part of Z represents the inductive reactance when $f > f_0$ because $X_C < X_L$.

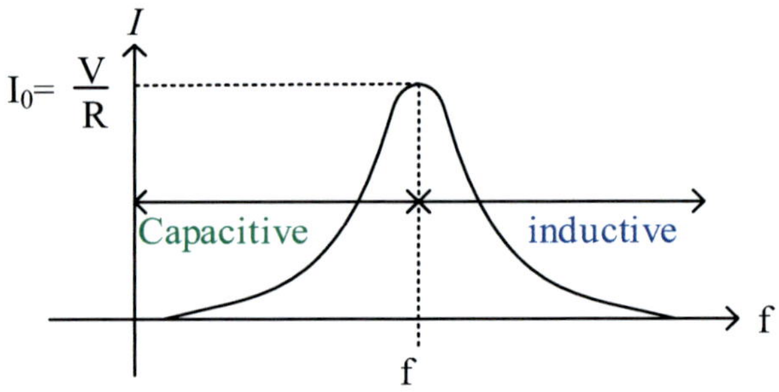

Fig. (4.8). *I-f* curve of series RLC circuit.

A series RLC circuit with impedance *Z* is illustrated in Fig. (4.9 **a**). In this figure, *V* and *I* represent the root mean square (rms) voltage and current, respectively. The impedance *Z* results in power consumption *S*. *S* is defined in equation 4.18. When $f \neq f_0$, current *I* produces an angle θ relative to *V*, as expressed in equation 4.19. The active power *P* is defined in equation 4.20. Fig. (4.9 **b**) displays the relationship among the apparent power (*S*), active power (*P*), and reactive power (*q*) in Cartesian coordinates. The aforementioned figure indicates that power consumption (*P*) only occurs in the real part (R) and decreases with an increase in θ.

$$S = VI \tag{4.18}$$

$$\theta = \tan^{-1}\frac{X_L - X_C}{R} \tag{4.19}$$

$$P = S\cos\theta = VI\cos\theta \tag{4.20}$$

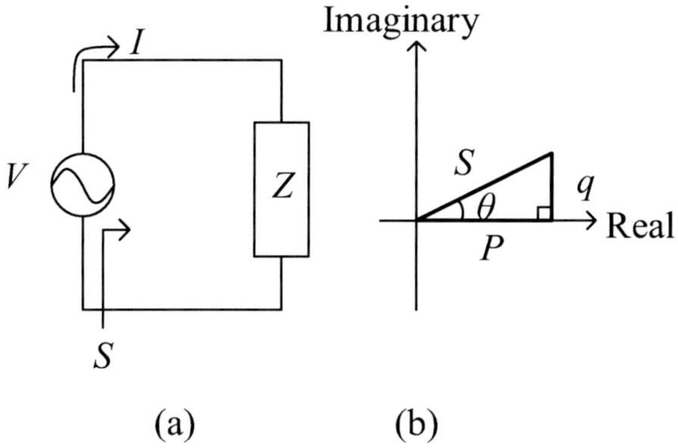

Fig. (4.9). Power consumption of *Z* in a series RLC circuit.

4.3.3. Quality factor

The *Q* factor of a series RLC circuit is expressed as follows:

$$Q = \frac{Q_{L0}}{P_0} = \frac{I^2 \times X_{L0}}{I^2 \times R} = \frac{X_{L0}}{R} = \frac{2\pi f_0 L}{R} \text{ or} \tag{4.21A}$$

$$Q = \frac{Q_{C0}}{P_0} = \frac{I^2 \times X_{C0}}{I^2 \times R} = \frac{X_{C0}}{R} = \frac{1}{2\pi f_0 RC} \tag{4.21 B}$$

The Equation (4.17) replace to (4.21) can obtain (4.22).

$$Q = \frac{2\pi L}{R} \times f_0 = \frac{2\pi L}{R} \times \frac{1}{2\pi\sqrt{LC}} = \frac{1}{R} \times \frac{L\sqrt{L}}{L\sqrt{C}} = \frac{1}{R} \times \sqrt{\frac{L}{C}} \quad \textbf{(4.22)}$$

In equation 4.22, the *Q* factor is not related to the frequency. The voltage of each component when resonance occurs in a series RLC circuit is expressed in equations 4.23 – 4.25.

$$V_{R0} = I_0 \times R = I_0 \times Z = V \quad \textbf{(4.23)}$$

$$V_{L0} = I_0 \times X_{L0} = \frac{V}{R} \times X_{L0} = \frac{X_{L0}}{R} \times V = Q \times V \quad \textbf{(4.24)}$$

$$V_{C0} = I_0 \times X_{C0} = \frac{V}{R} \times X_{C0} = \frac{X_{C0}}{R} \times V = Q \times V \quad \textbf{(4.25)}$$

If Q > 1 and resonance occurs, the voltage of the inductor and capacitor is higher than the source voltage v_{in}, as expressed in equations 4.24 and 4.25.

The voltage gain Gjω for the RLC circuit displayed in Fig. (**4.6** is derived using equations 4.26 and 4.27. Equation 4.28 presents the relationship between the phaseω and the frequency.

$$G(j\omega) = \frac{v_L}{v_{in}} = \frac{i_{in} \times j\omega L}{i_{in} \times Z} = \frac{j\omega L}{R + j\left(\omega L - \frac{1}{\omega C}\right)} = \frac{1}{1 - \frac{1}{\omega^2 LC} - j\frac{R}{\omega L}} \quad \textbf{(4.26)}$$

$$= \frac{1}{\frac{1-\omega^2 LC}{\omega^2 LC} - j\frac{R}{\omega L}} \quad \textbf{(4.27)}$$

$$\text{Phase}(\omega) = -\tan^{-1}\frac{-\frac{R}{\omega L}}{\frac{1-\omega^2 LC}{\omega^2 LC}} \quad \textbf{(4.28)}$$

The relationships of the voltage gain and phase with the relative frequency can be represented using Bode diagrams (Fig. **4.10**). In Fig. (**4.10**), at f_0, the inductor's voltage is $2 \times v_{in}$ when $Q = 2$. Moreover, the inductor voltage is $10 \times v_{in}$ when $Q = 10$, and the phase change is steeper when $Q = 10$ than when $Q = 2$.

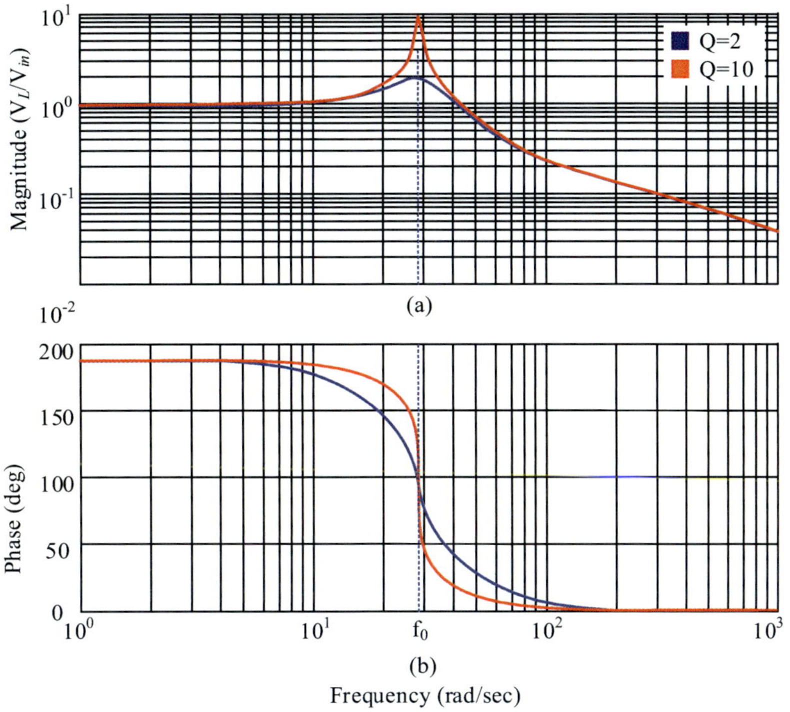

Fig. (4.10). Frequency responses of a series RLC circuit with changes in the voltage gain and phase of X_L.

4.3.4. Bandwidth and Selectivity

The *I–f* curve of a series RLC circuit, displayed in Fig. (**4.11**), indicates that the maximum current I_0 occurs at f_0. In Fig. (**4.1**), the BW is between f0' and f0". If $Q \geq 10$, the *I–f* curve between f0' and f0" is symmetric. The BW under the aforementioned condition is expressed in equation 4.29.

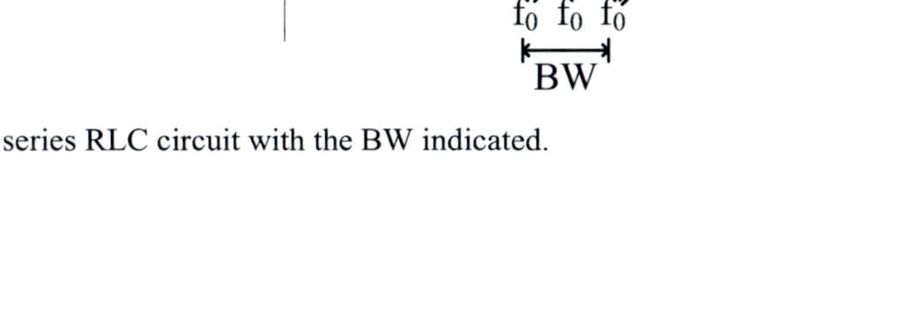

Fig. (4.11). *I-f* curve of a series RLC circuit with the BW indicated.

$$\mathrm{BW} = \frac{f_0}{Q} = \frac{\frac{1}{2\pi\sqrt{LC}}}{\frac{1}{R}\times\sqrt{\frac{L}{C}}} = \frac{R}{2\pi L} \qquad \textbf{(4.29)}$$

A sharper *I–f* curve within the BW indicates better frequency selectivity. Conversely, a flatter *I–f* curve within the BW indicates poorer frequency selectivity. Thus, the *Q* factor affects the BW and *I–f* curve of a series RLC circuit as follows:

- The higher the value of *Q*, the smaller the BW, the sharper the *I–f* curve, and the higher the frequency selectivity.
- The smaller the value of *Q*, the higher the BW, the flatter the *I–f* curve, and the worse the frequency selectivity.

The parameters *R* and $\frac{L}{C}$ from equation 4.22 affect the *Q* factor and BW of a series RLC circuit as follows:

- If *R* is constant, *Q* increases and the BW decreases when $\frac{L}{C}$ increases (Fig. **4.12**).
- If $\frac{L}{C}$ is constant, *Q* decreases, *I* decreases, and the BW increases when *R* increases (Fig. **4.13**).

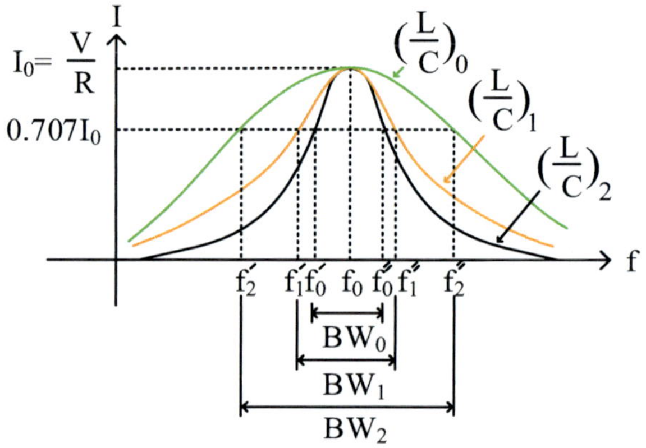

Fig. (4.12). *I –f* curve of a series RLC circuit when R is constant and $\left(\frac{L}{C}\right)_0 < \left(\frac{L}{C}\right)_1 < \left(\frac{L}{C}\right)_2$

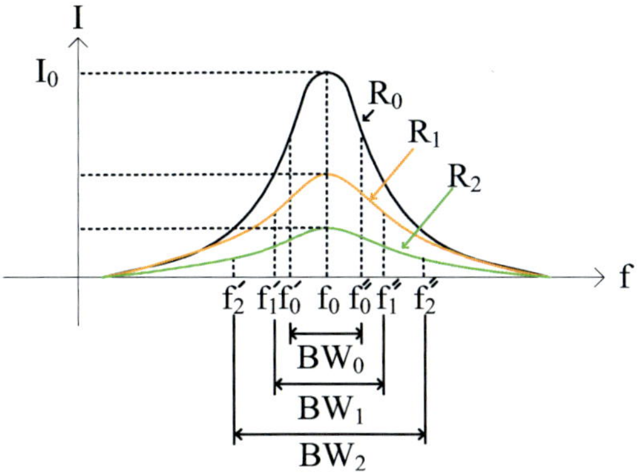

Fig. (4.13). I –f curve of a series RLC circuit when LC is constant and $R_0 < R_1 < R_2$.

4.4. Parallel RLC resonant circuit

In a parallel RLC circuit, the impedance Z is replaced by the admittance Y ($Y = Z^{-1}$), the resistance R is replaced by the electrical conductance G ($G = R^{-1}$), and the reactance X is replaced by the susceptance B ($B = X^{-1}$) for easy calculation. A parallel RLC circuit is displayed in Fig. (**4.14**). The admittance Y of this circuit is expressed in equation 4.30. When the frequency of the source voltage is f_0, resonance occurs in the aforementioned circuit, and this resonance results in a minimum Y value, as presented in equation 4.31. The resonance frequency f_0 is expressed in equation 4.32 and 4.33. The relationships of B_L, B_C, and Y with the source voltage's frequency are illustrated in Fig. (**4.15**), in which B_{L0} and B_{C0} represent the inductive and capacitive susceptance at f_0, respectively.

$$Y = G + j\,B_C - j\,B_L \tag{4.30}$$

$$\text{Min}\,Y = G + j\left(2\,\pi\,f_0 C - \frac{1}{2\,\pi\,f_0 L}\right) = G \tag{4.31}$$

$$\because B_C = B_L \text{ at f0}$$

$$\therefore 2\,\pi\,f_0 C = \frac{1}{2\,\pi\,f_0 L} \tag{4.32}$$

$$f_0 = \frac{1}{2\,\pi\,\sqrt{LC}} \tag{4.33}$$

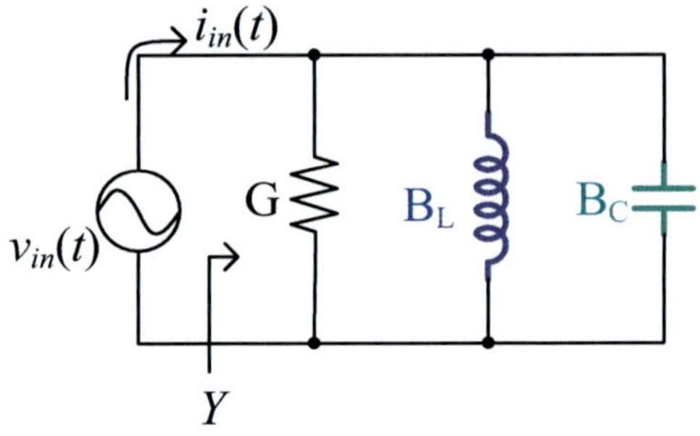

Fig. (4.14). Parallel RLC resonant circuit.

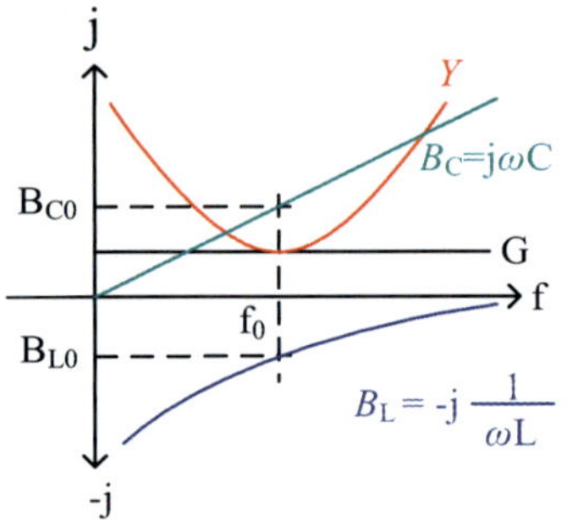

Fig. (4.15). Relationships of B_C, B_L, and Y with source voltage's frequency.

4.4.1. Power

The *I–f* curve of a parallel RLC circuit is displayed in Fig. (**4.16**). In this figure, the imaginary part of the admittance Y represents the inductive susceptance when $f < f_0$ because $B_C < B_L$ under this condition. The imaginary part of Y is 0 when $f = f_0$ because $B_C = B_L$ under this condition. The minimum current also occurs when $f = f_0$. The imaginary part of Y represents the capacitive susceptance when $f > f_0$ because $B_C > B_L$ under this condition.

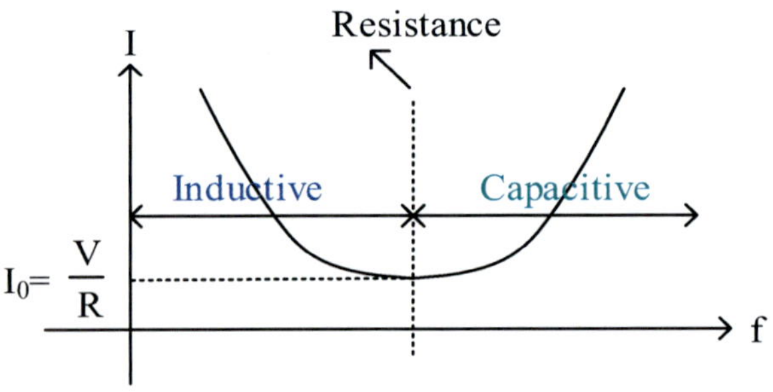

Fig. (4.16). *I–f* curve of a parallel RLC circuit.

Fig. (**4.17a**) displays a parallel RLC circuit with admittance *Y*. In this figure, *V* and *I* represent the rms voltage and current, respectively. The admittance *Y* results in power consumption *S*. *S* is defined in equation 4.34. When $f \neq f_0$, the current *I* flowing through the resistor produces an angle θ relative to *V*, as expressed in equation 4.35. The active power *P* is defined in equation 4.36. The relationship among the apparent power (*S*), active power (*P*), and reactive power (*q*) is illustrated in Cartesian coordinates in Fig. (**4.17b**), meaning that S is equal to the square of (P^2+q^2).

$$S = V\,I \quad \textbf{(4.34)}$$

$$\theta = tan^{-1}\frac{B_L - B_C}{G} \quad \textbf{(4.35)}$$

$$P = S\cos\theta = VI\cos\theta \quad \textbf{(4.36)}$$

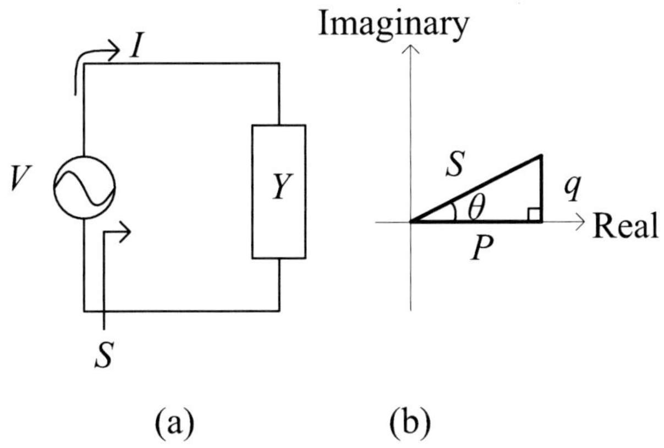

Fig. (4.17). Power consumption caused by *Y*.

4.4.2. Quality factor

The *Q* factor of a parallel RLC circuit is expressed in equations 4.37a and 4.37b.

$$Q = \frac{Q_{L0}}{P_0} = \frac{V^2 \times B_{L0}}{V^2 \times G} = \frac{B_{L0}}{G} = \frac{R}{X_{L0}} = \frac{R}{2\pi f_0 L} \text{ or} \quad \textbf{(4.37 A)}$$

$$Q = \frac{Q_{C0}}{P_0} = \frac{V^2 \times B_{C0}}{V^2 \times G} = \frac{B_{C0}}{G} = \frac{R}{X_{C0}} = 2\pi f_0 CR \quad \textbf{(4.37 B)}$$

Equation 4.33 is substituted into 4-37a to obtain the following equation:

$$Q = \frac{R}{2\pi L} \times \frac{1}{f_0} = \frac{R}{2\pi L} \times \frac{1}{\frac{1}{2\pi\sqrt{LC}}}$$

$$Q = R \times \sqrt{\frac{C}{L}} \qquad \textbf{(4.38)}$$

In equation 4.38, the *Q* factor is not related to the frequency. The current of each component when resonance occurs in a parallel RLC circuit is expressed in equations 4.39 –4-41.

$$I_{R0} = V \times G = V \times Y_0 = I_0 \qquad \textbf{(4.39)}$$

$$I_{L0} = V \times B_{L0} = \frac{I_0}{G} \times B_{L0} = \frac{B_{L0}}{G} \times I_0 = Q \times I_0 \qquad \textbf{(4.40)}$$

$$I_{C0} = V \times B_{C0} = \frac{I_0}{G} \times B_{C0} = \frac{B_{C0}}{G} \times I_0 = Q \times I_0 \qquad \textbf{(4.41)}$$

If Q > 1 and resonance occurs, the current of the inductor and the voltage of the capacitor is higher than source current I_0, as expressed in equations 4.40 and 4.41.

4.4.3. BW and Selectivity

The *I–f* curve of a parallel RLC circuit, displayed in Fig. (**4.18**), indicates that the minimum current I_0 occurs at f_0. In this figure, the BW is between f0' and f0''. If $Q \geq 10$, the curve between f0' and f0'' is symmetric. The BW under the aforementioned condition is expressed in equation 4.42.

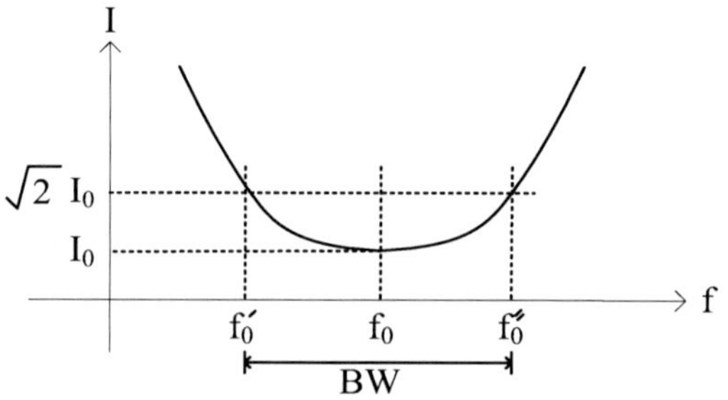

Fig. (4.18). *I–f* curve of a parallel RLC circuit with the BW indicated.

$$BW = \frac{f_0}{Q} = \frac{\frac{1}{2\pi\sqrt{LC}}}{R \times \sqrt{\frac{C}{L}}} = \frac{1}{2\pi RC} \qquad \textbf{(4.42)}$$

A sharper *I–f* curve within the BW indicates higher frequency selectivity. Conversely, a flatter *I–f* curve within the BW indicates poorer frequency selectivity. Therefore, the *Q* factor affects the BW and *I–f* curve of a parallel RLC circuit as follows:

- The higher the value of *Q*, the smaller the BW, the sharper the *I–f* curve, and the higher the frequency selectivity.

The parameters *R* and $\frac{C}{L}$ from equation 4.38 affect the *Q* factor and BW of a parallel RLC circuit as follows:

- If *R* is constant, *Q* increases, and the BW decreases when $\frac{C}{L}$ increases (Fig. **4.19**).
- If $\frac{C}{L}$ is constant, *Q* decreases, I_0 increases, and the BW increases when *R* decreases (Fig. **4.20**).

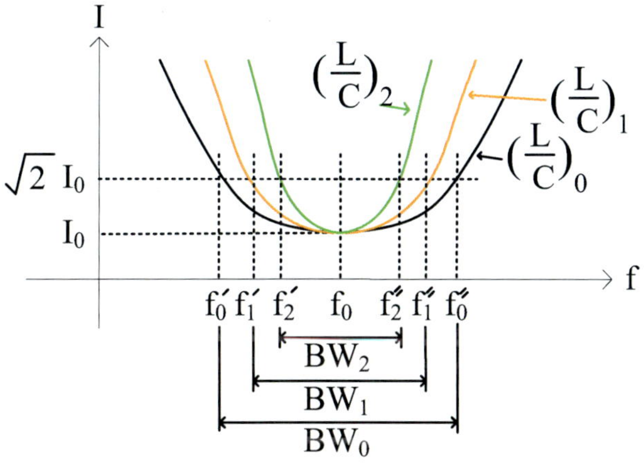

Fig. (4.19). *I–f* curve of a parallel RLC circuit when *R* is constant and $\left(\frac{L}{C}\right)_0 < \left(\frac{L}{C}\right)_1 < \left(\frac{L}{C}\right)_2$

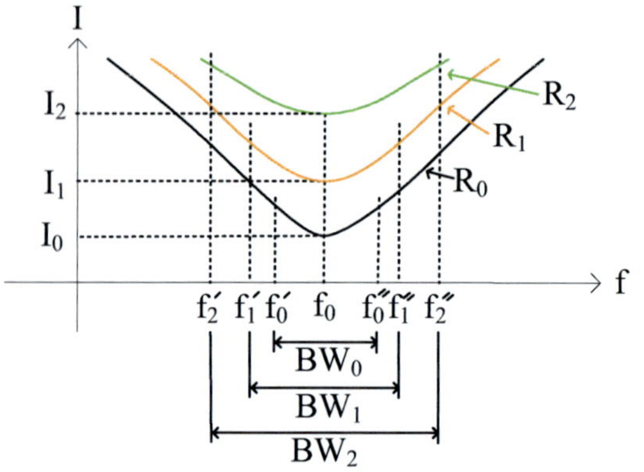

Fig. (4.20). *I–f* curve of a parallel RLC circuit when CL is constant and $R_0 > R_1 > R_2$.

CONCLUSION

Two types of LC networks are typically used in battery charging systems: series and parallel LC networks. In this chapter, series and parallel LC networks are described using series and parallel RLC resonant circuits, respectively. A series LC resonant network has minimal impedance, high voltage gain, which is proportional to the *Q* factor and a BW that is inversely proportional to the *Q* factor. A parallel LC resonant network has minimal admittance, high current gain, which is proportional to the *Q* factor, and a BW that is inversely proportional to the *Q* factor. This chapter presents the operational principles of the aforementioned two LC networks, which are the foundation for developing complex tuned networks, such as LCL and LCC networks.

CHAPTER 5

Inductive and Magnetic Resonance

Abstract: Wireless inductive power transfer can enhance user convenience and safety in the charging of electric vehicle batteries. This method can be used for transmitting high power through an air gap of 0.01–0.2 m at the efficiency of up to 90%, a quality factor (Q factor) of approximately 5–100, a coupling factor of approximately 0.2–0.5, and a fundamental operating frequency of hundreds of hertz to several hundred kilohertz. A longer wireless power transmission distance can be achieved by using resonant tanks with magnetic coupling in which the source and load are not in contact than can be achieved through wireless inductive power transfer. With the aforementioned tanks, efficient wireless power transmission can be achieved with a Q factor of approximately 100–10 000, a coupling factor of less than 0.2, and an operating frequency range of 0.2 to 20 MHz. The aforementioned two methods are based on the principle of electromagnetic induction. Implementation of these methods requires the use of advanced semiconductors, high-frequency switching components with low power loss, a tuned compensation network that can tolerate lateral misalignment between coils, and a power pad design that enables high magnetic flux to be generated for transmission over a relatively long distance.

Keywords: Compensation network, Electric vehicle (EV) battery charging, High-frequency switching components, Inductive power transfer, Magnetic resonance coupling.

1.1. INTRODUCTION

Wireless power transfer (WPT), including inductive power transfer (IPT) and magnetic resonance coupling (MRC; *i.e.*, loose coupling), has attracted considerable research attention over the past decade because of its convenience and security. WPT is based on the principle of electromagnetic induction [178, 190]. In WPT, the air gap is smaller than λ, the near-field transmission [167, 191, 192]. IPT has developed quickly in the last decade [121, 192, 193]. In particular, considerable research and development efforts have been focused on the charging of electric vehicle (EV) batteries [102, 194, 195]. Traffic systems that can be charged through static or dynamic wireless charging, including those for road vehicles and trains, have been developed. The output power levels of these systems range from a few kilowatts to the megawatt range [179, 196 - 198]. Achieving the efficient wireless transmission of considerable power betw-

Chih-Cheng Huang & Chun-Liang Lin

een the transmission (TX) coil in a primary tank and the receiver (RX) coil in a secondary tank through the air gap between them is critical. Primary and secondary tanks contain a coil and a compensated network, which can be tuned to achieve voltage or current gain.

Four basic types of compensation network exist: series–series (SS), series–parallel (SP), parallel–series (PS), and parallel–parallel (PP) compensation networks. These networks have been analyzed in detail and compared [112, 199, 200]. Complex compensation networks that include a combination of SP inductor–capacitor (LC) circuits, such as inductor–capacitor–inductor (LCL) and inductor–capacitor–capacitor (LCC) networks, have also been developed [185, 201 - 203]. The coupling factor is the main limiting variable in dynamic charging applications. The load conditions can be considered to be invariant during the transient charging process [198, 204 - 206]. An EV battery charger with an LCL primary tank and a parallel LC tank of the secondary side can transmit 3 kW of power at efficiency of approximately 92% over a distance of up to three times the radius of its coils under an operating frequency of 30 kHz and air gap of 16 cm [182].

An MRC system can wirelessly transmit power over a long distance under high operating frequency. Its primary and secondary tanks have the same resonant frequency, with these tanks containing an isolated source and load, respectively. The prototype MRC system designed in [103], which generates strongly coupled magnetic resonance, can transmit 60 W of power with efficiency of approximately 40% over a distance of more than 2 m when the frequency of the source voltage is 9.9 MHz [103]. This prototype exhibits efficient nonradiative power transfer over distances of up to eight times the radius of its coils and more than three times the radius of an IPT system's coils.

In this chapter, we introduce the power pad design of LCL and LCC compensated networks and examine how an MRC system's transmission distance is related to its quality factor (Q factor), coupling factor k, and resonant angular frequency.

1.2. Inductive Resonance

The principle of electromagnetic induction is based on Ampere's law and Faraday's law. In Fig. (5.1 **a**), the circuit on the left side, which has voltage U_1 and current i_1, generates magnetic field B around coil L_1 as per (5-1). The parameter μ0 is the permeability of vacuum, and r is the radius of $\vec{B}$ (Fig. 5.1 **b**). Equation (5.2) indicates that the current in the aforementioned circuit generates magnetic flux Φ_B. The magnetic flux close to the current is stronger [red circles in Fig. (**5.1b**)] than that away from the current [pink circles in Fig. (**5.1b**)].

The coil L_2 in the circuit on the right side of Fig. (5.1 **a**) generates a magnetic field ($-B$) that is equal in magnitude but opposite in direction to the time-varying magnetic field B. Therefore, voltage U_2 is generated in L_2 by $-B$ as per (5.3), in which N is the number of turns and Φ_B is the magnetic flux. The voltage U_2 is proportional to Φ_B. Higher current i_2 corresponds to higher U_2 [Φ_B is proportional to i_2, as presented in Figs. (**5.1** and **5.2**)]. The magnetic flux Φ_B causes the coupling of L_1 and L_2, which results in power being transferred from the circuit on the left side of Fig. (**5.1a**) to the circuit on the right side of Fig. (**5.1a**) through the air gap.

$$B = \mu_0 \frac{i_1}{2\pi r} \tag{5.1}$$

$$\Phi_B = \iint B \cdot dA \tag{5.2}$$

$$U_2 = -N \frac{d\Phi_B}{dt} \tag{5.3}$$

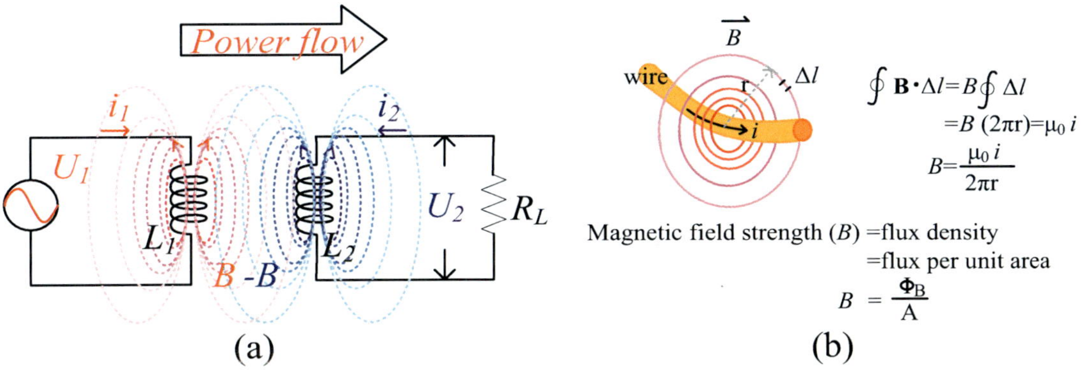

Fig. (5.1). Fundamental WPT based on Ampère's circuital law and Faraday's law of Faraday's low of electromagnetic induction.

To achieve the high-efficiency transmission of high power, a resonant tank that can achieve voltage or current gain is added to a compensated network. Such a tank can strengthen the electromagnetic effect and increase the inductive distance (Fig. **5.2**). In the aforementioned network, a magnetic field generates an eddy current and causes iron loss, whereas a current generates resistance in the coil and causes copper loss. The aforementioned phenomena are not shown in Fig. (**5.2**) for convenience. U_1 represents the high-frequency voltage generated by a class E amplifier, which is used in insulated-gate bipolar transistors and metal–oxide–semiconductor field-effect transistors. In loose coupling applications, the coupling factor of electromagnetic induction is approximately 0.2–0.5 depending on the size and shape of the power pad. For example, if the power pad's diameter is increased, the coupling factor k also increases. The coupling factor is expressed in (5.4). In Fig. (**5.2**), reflected voltages U_{21} and U_{12} are generated by the mutual inductance L_m. The ferrite (Fe) mounting under the coil can guide magnetic flux to considerably limit electromagnetic radiation. The impedance of the secondary side of the WPT system displayed in Fig. (**5.2**) (denoted Z_S) is expressed in (5.5). Equation (5.5) is substituted into (5.6) to obtain the impedance Z_{21}.

$$\mathrm{k} = \frac{L_m}{\sqrt{(L_1 L_2)}} \tag{5.4}$$

$$Z_S = \frac{j\omega L_m \cdot i_1}{i_2} \tag{5.5}$$

$$Z_{21} = \frac{U_{21}}{i_1} = \frac{-j\omega L_m}{i_1} \cdot \frac{j\omega L_m \cdot i_1}{Z_S} = \frac{(\omega L_m)^2}{Z_S} \tag{5.6}$$

The power transmitted from the primary side of the aforementioned WPT system to its secondary side is as follows:

$$\mathrm{P} = (Re\, Z_{21}) i_1^2 \tag{5.7}$$

where "*Re*" represents the real part of the impedance Z_{21}.

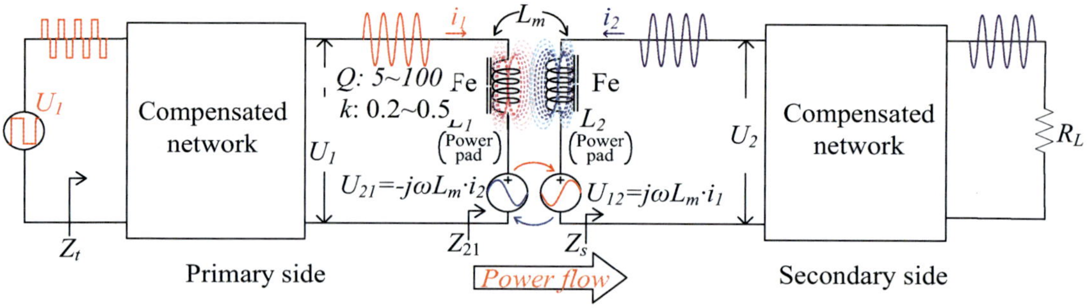

Fig. (5.2). General two-coil WPT system.

1.2.1 Basic Compensated Topologies

A compensated network enhances the efficiency of WPT. The four basic compensated network topologies are illustrated in Fig. (**5.3**). In this figure, the first letter represents the type of compensated network in the primary tank, and the second letter represents the type of compensated network in the secondary tank. For example, in Fig. (5.3 **b**), the primary and secondary tanks are a series LC network and parallel LC network, respectively. The fundamental parameters of series and parallel LC compensation networks are listed in Table **5.1**. The resonant frequency of the primary tank (primary compensated network) is expressed as follows:

$$f_0 = \frac{1}{2\pi\sqrt{L_1 C_1}} \tag{5.8}$$

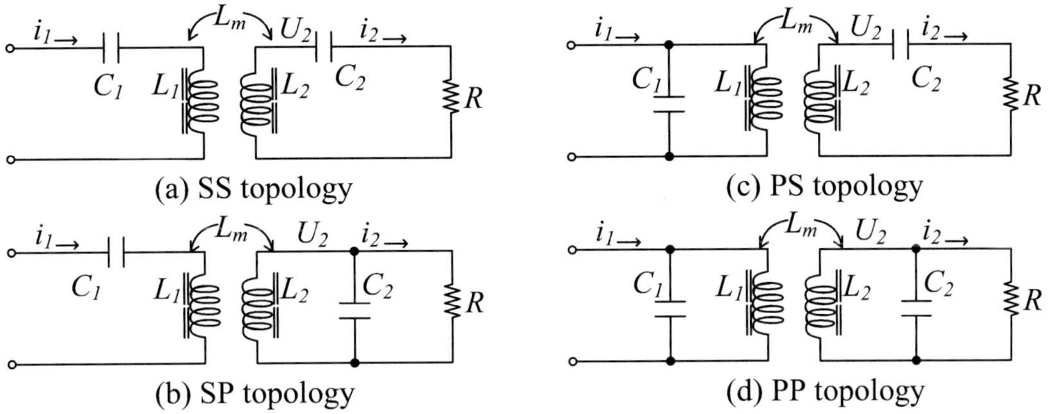

Fig. (5.3). Basic compensated network topologies.

Table 1. Fundamental parameters of series and parallel lc compensated networks.

Item	Series LC	Parallel LC
Z_S	$j\omega L_2 + \frac{1}{j\omega C_2} + R_L$	$j\omega L_2 + \frac{1}{j\omega C_2 + \frac{1}{R_L}}$
Load current	i_2	$\frac{U_2}{R_L}$
Load voltage	$i_2 R_L$	U_2
Quality factor Q_2	$\frac{\omega_0 L_2}{R_L}$	$\frac{R_L}{\omega_0 L_2}$
Reflected impedance Z_{21}	$\frac{(\omega_0 L_m)^2}{R_L}$	$\frac{R L_m^2}{L_2^2} - j\frac{\omega_0 L_m^2}{L_2}$
Z_t	$j\omega L_1 + \frac{1}{j\omega C_1} + Z_{21}$	$\frac{1}{j\omega C_1 + \frac{1}{j\omega L_1 + Z_{21}}}$

At coil L_1, the apparent power S_1 is equal to $U_1 \times i_1$, and the active power is equal to $U_1 \times i_1 \times \cos\theta$. To reduce the amount of power loss, a minimum Volt-Ampere (VA) rating is required. Therefore, $S_1 = P_1$ and $\theta = 0°$ must be achieved in the primary tank. These conditions are called the zero-phase-angle (ZPA) conditions [200, 207]. In the initial stage of designing an EV battery charger, the parameters of L_1—including the shape and size of the power pad, inductance, ferrite content, and Q factor—must be determined. To achieve ZPA conditions in the primary tank, C_1 must complement L_1 and Z_{21} (Fig. **5.2**). When C_1 complements L_1 and Z_{21}, the primary resonant tank exhibits resonance. For example, the WPT architecture using SS and the C_1 as (5-9), if C_1=C_2 and L_1=L_2, then the primary tank and secondary tank is the same resonant frequency. The C_1 is as Eq. (5-10) without the source's frequency.

$$C_1 = \frac{1}{\omega^2 L_1} \tag{5.9}$$

$$C_1 = \frac{L_2 C_2}{L_1} \tag{5.10}$$

Table **5.2** presents the equations for determining C_1 for the four basic compensated topologies when ZPA conditions are achieved. In the SS topology, C_1 is independent of the coupling factor k and load condition. In the SP topology, C_1 varies with the coupling factor. In the PS and PP topologies, C_1 is affected by the coupling factor and load conditions (Q factor).

Table 2. Equations of the Compensated Capacitance of the Primary Tank (C_1) when the zpa Conditions are Satisfied.

Topology	C_1
SS	$\frac{L_2C_2}{L_1}$
SP	$\frac{L_2C_2}{L_1} \times \frac{1}{1-k^2}$
PS	$\frac{L_2C_2}{L_1} \times \frac{1}{Q_2^2k^4+1}$
PP	$\frac{L_2C_2}{L_1} \times \frac{1-k^2}{Q_2^2k^4+(1-k^2)}$

1.2.2 LCL/LCC-compensated Topologies

When $Q > 1$, a series resistor–inductor–capacitor (RLC) circuit amplifies voltage, whereas a parallel RLC circuit amplifies current (Chapter 4 of this e-book). A resonant series LC circuit amplifies voltage, whereas a parallel resonant LC circuit amplifies current.

A WPT system in which the primary tank contains both a series LC circuit and parallel LC circuit is called a hybrid SP (LCL) compensated network. Tuned LCL compensated networks are widely used in EV battery charging. At the resonant frequency, LCL networks function similar to a current source (*i.e.*, a parallel LC circuit). Under this condition, the primary-side coil is controlled by the high-frequency square-wave voltage supplied by the power inverter, regardless of the coupling factor k and load of the secondary side; thus, the control mechanism in the primary tanks is easy by tuning the LCL parameters, the reactive power is fully compensated and the power inverter provides only active power; thus, the VA rating required for the inverter output power can be minimized.

LCC compensation networks, which are based on the LCL compensated network, have been successfully used for stationary wireless charging. LCC compensation networks facilitate zero-voltage switching [208, 209] or zero-current switching [210, 211] by the inverter; increase tolerance of lateral misalignment [212, 213]; and increase the WPT efficiency. Fig. (**5.4**) illustrates a typical WPT system for EV battery charging that contains an LCC or LCL compensated network on its primary side and a parallel LC compensated network on its secondary side.

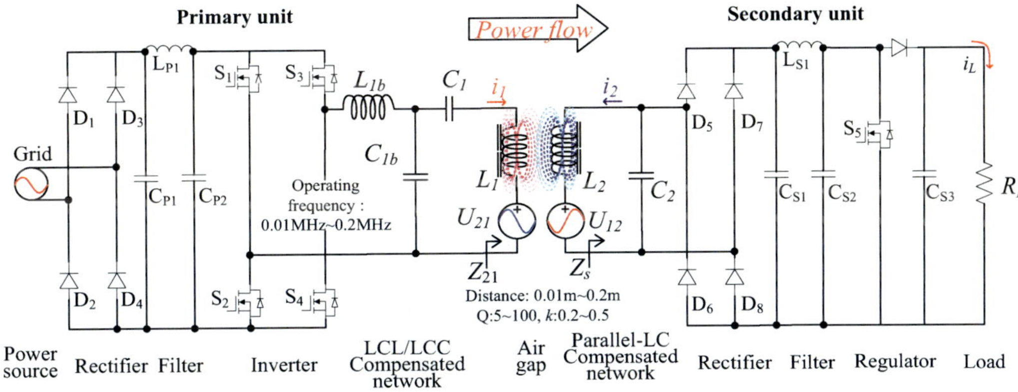

Fig. (5.4). Typical EV battery charging system with an LCC compensated network on its primary side and a parallel LC compensated network on its secondary side.

On the primary side of the system displayed in Fig. (**5.4**), a full-bridge inverter outputs a high-frequency square voltage to an LCL or LCC compensated network to link the primary tank to the secondary tank through i_1 and i_2 as per (5-3). The secondary tank (with inductance L_2 and capacitance C_2) acts similarly to a current source that generates constant current in L_{S1}. The output power of the system displayed in Fig. (**5.4**) can be controlled by changing the duty ratio of switch S_5. Use of an optimized LCC compensated network in the aforementioned system can provide a high degree of design freedom and achieve stable power being generated over a wide range of coupling factors [185]. The output power properties of different topologies are illustrated in Fig. (**5.5**). The output power in the aforementioned figure is normalized using the rated power. Fig. (**5.5**) indicates that the optimal LCC topology enables stable power generation.

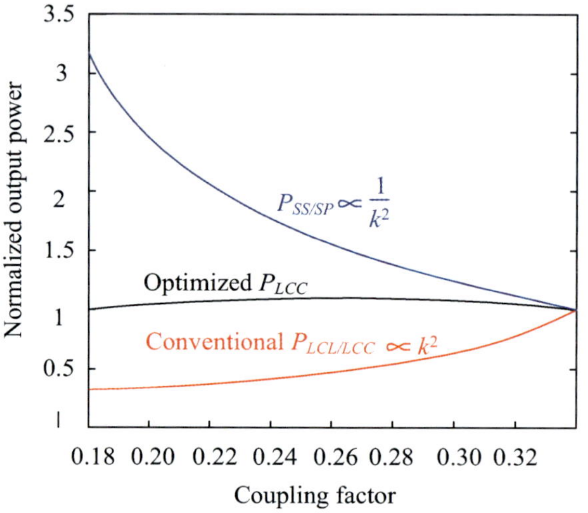

Fig. (5.5). Power profile of different compensated network topologies.

1.2.3 Power Pad

A circular power transmission pad [214, 215] with magnetic structures is used to couple the flux between L_1 and L_2 through an air gap. A power transmission pad contains four major components (Fig. **5.6**): a winding, housing, ferrites, and an aluminum backing plate. The aluminum backing plate under the ferrite provides shielding and allows heat loss. To increase the air gap between L_1 and L_2, a Double D shape (DD) type power pad has been developed [114]. The DD-type power pad is shown in Fig. (5.7 **a**). As depicted in this figure, the pad generates a single-sided magnetic field, the height of which is approximately half the diameter of the pad. The mainly magnetic height of the DD-type power pad is two times of the circular type pad.

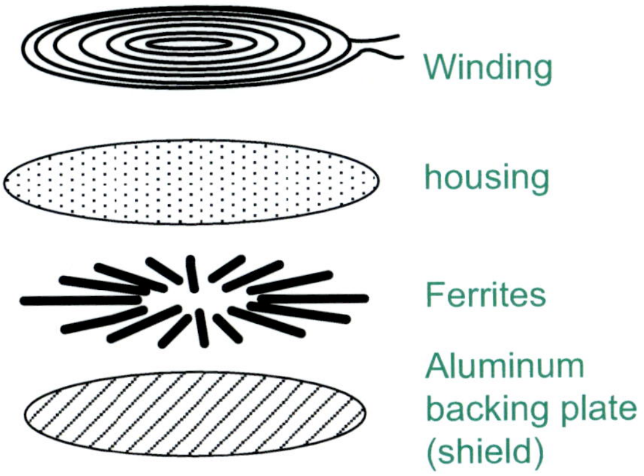

Fig. (5.6). Structure of a power transmission pad.

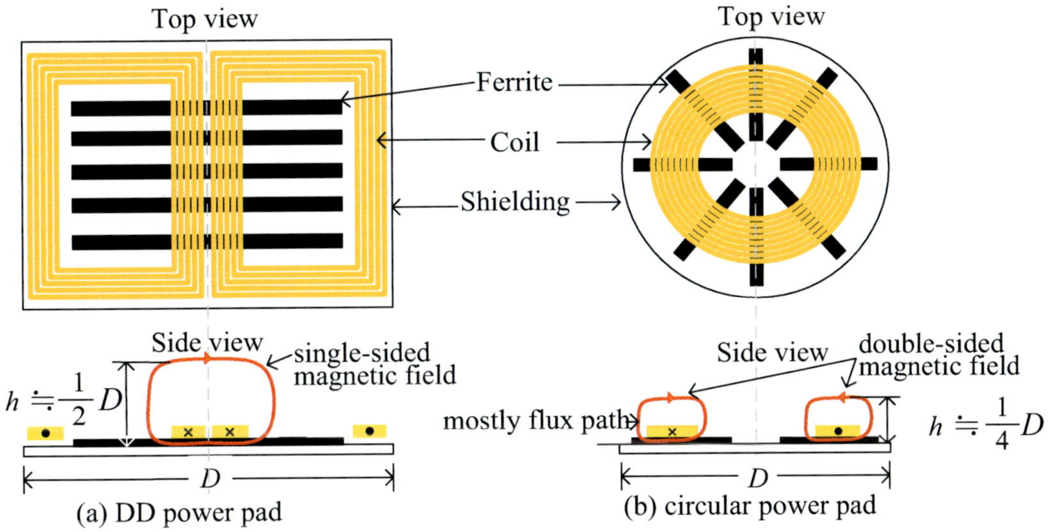

Fig. (5.7). Two typical types of power pad.

1.3. Magnetic Resonance

The advantage of IPT is that it enables the efficient wireless transfer of high power over an air gap through ferrite, which guides the magnetic flux. However, the disadvantage of an IPT system is that its air gap is only approximately 0.01–0.2 m. An MRC system contains coils without ferrite that can wirelessly transfer nonradiated power with acceptable efficiency over a greater air gap (approximately 0.1–2.0 m) than can IPT systems.

The WPT in IPT and MRC systems involves the generation of a nonradiated inductive electromagnetic field. A typical MRC system is illustrated in Fig. (**5.8**). In this figure, the inverter output frequency is 0.2–20 MHz. The source tank contains a series LC compensated network with capacitance C_{1b} and inductance L_{1b}. The TX and RX tanks contain a series LC compensated network with inductance L and capacitance C. Because the ai1r gap between the source and TX tanks is small, high power density and transmission efficiency are achieved for transmission between these tanks. The aforementioned tanks are isolated from each other to maintain a high Q factor and resonant angular frequency ω_0.

The TX tank amplifies the resonant voltage to generate a strong electromagnetic field for increasing the air gap. Both LC tanks have extremely high Q factors and the same resonant angular frequency ω_0 without ferrite is called MRC in most papers. A longer air gap is present in an MRC system than in an IPT system. One reason for this is that the resonant frequency of an MRC system is higher than that of an IPT system. Because a higher angular frequency results in more rapidly varying magnetic flux Φ_B, the RX tank's inductive voltage is higher as per (5-5)–(5-11). High power density and transmission efficiency are achieved for transmission between the RX and load tanks. Finally, the output power is absorbed by the load through the regulator. Because of the large air gap and small inductance L_m between the TX and RX tanks, the reflected voltages U_{21} and U_{12} and reflected impedance Z_{21} (Fig. **5.4**) can be ignored. To provide a simplified representation, the copper losses in the source, TX, RX, and load tanks are ignored in Fig. (**5.8**).

$$V_{RX\ tank} = -\frac{d\Phi_B}{dt} \quad \textbf{(5.11)}$$

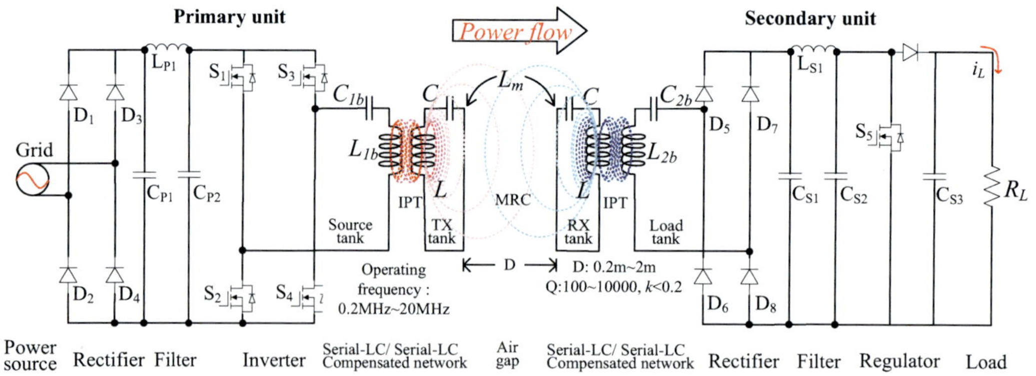

Fig. (5.8). A typical long transmission-distance WPT system with MRC.

Equation (5.12) presents the expression for the reactance of the equivalent circuit at resonance, which results in two frequency-splitting phenomena. These phenomena are expressed in (5.13) and (5.14) [216]. The coupling factor k can be derived using (5.13) and (5.14) and substituted into (5.15).

$$\frac{1}{\omega L_m} + \frac{2}{\omega(L-L_m)-\frac{1}{\omega C}} = 0 \quad \textbf{(5.12)}$$

$$\omega_m = \frac{\omega_0}{\sqrt{1+k}} = \frac{1}{\sqrt{(1+L_m)C}} \quad \textbf{(5.13)}$$

$$\omega_e = \frac{\omega_0}{\sqrt{1-k}} = \frac{1}{\left(\sqrt{L-L_m}\right)C} \quad \textbf{(5.14)}$$

$$k = \frac{L_m}{L} = \frac{\omega_e^2-\omega_m^2}{\omega_e^2+\omega_m^2} \quad \textbf{(5.15)}$$

An MRC system's efficiency is calculated using (5-16) and (5-17), where S_{11} and S_{21} represent the wave reflection and transmission ratios, respectively. The parameter S_{21} is calculated using (5-18) [217].

$$\eta_{11} = |S_{11}|^2 \times 100\% \quad \textbf{(5.16)}$$

$$\eta_{21} = |S_{21}|^2 \times 100\% \quad \textbf{(5.17)}$$

$$S_{21}(\omega) = \frac{2jL_m Z_0 \omega}{L_m^2\omega^2+\left[(Z_0+R)j\left(\omega L-\frac{1}{\omega C}\right)\right]^2} \quad \textbf{(5.18)}$$

An MRC system's efficiency is illustrated in Fig. (**5.9**). In Figs. (**5.9a** and **5.9b**), the air gap is within the range for overcoupling [218, 219]. In these figures, efficiency peaks are observed at ω_m and ω_e as per (5-13) and (5-14), and an efficiency trough is observed at ω_0. To increase the efficiency at ω_0 for an air gap within the overcoupling region, impedance matching systems have been proposed in [216, 220].

The difference between ω_m and ω_e can be decreased by increasing the air gap. The locations ω_m, ω_e, and ω_0 combine to become ω_0 upon critical coupling, as shown in Fig. (**5.9 c**). The efficiency η_{21} is almost the same at ω_m and ω_e when the air gap is in the overcoupling region. Finally, when the air gap is further increased, η_{21} decays and then stabilizes when the air gap is within the undercoupling region.

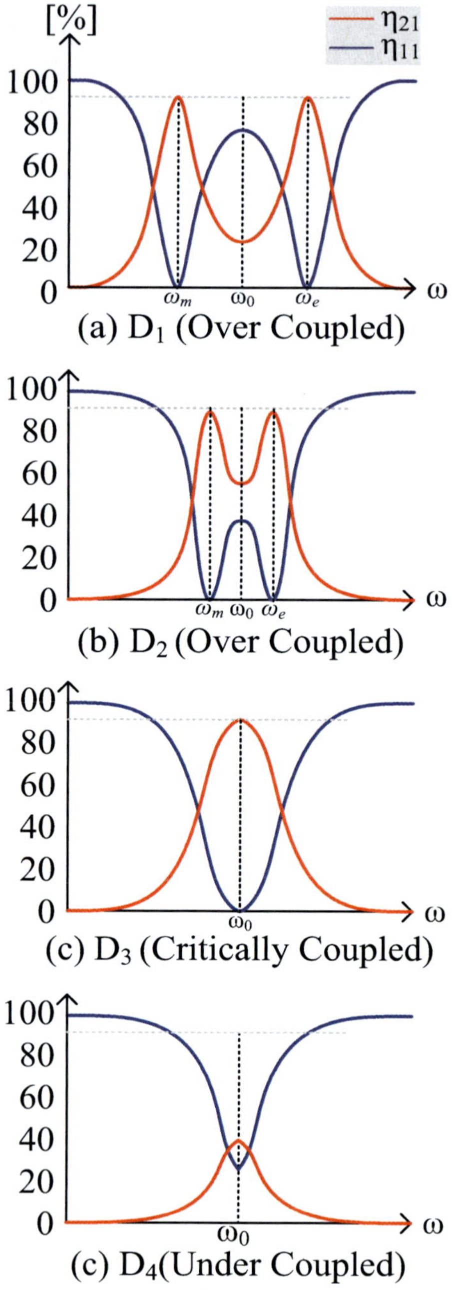

Fig. (5.9). Experimental results for the variation in transmission efficiency with ω under different air gaps ($D_1 < D_2 < D_3 < D_4$).

As illustrated in Fig. (**5.9**), efficient wireless transmission of high power over an acceptable air gap can be achieved using IPT and MRC systems. Future studies can combine IPT with MRC to achieve the efficient wireless transfer of high power over a large air gap.

CONCLUSION

WPT, including IPT and MRC, is based on the principle of electromagnetic induction. Rapid advances have been achieved in IPT over the last decade. IPT systems for EV battery charging have an efficiency of up to 90%, a Q factor of approximately 5–100, a coupling factor of approximately 0.2–0.5, an operating frequency ranging from hundreds of hertz to several hundred kilohertz, and an air gap of 0.01–0.2 m. The advantages of these systems include their compactness, ferrite-containing power pad that produces a low-leakage magnetic field, and simplicity for mass production. A disadvantage of the aforementioned systems is their small air gap.

An MRC system has efficiency of approximately 40%, a Q factor of approximately 100–10000, a coupling factor of less than 0.2, an operating frequency range of approximately 0.2–20 MHz, and an air gap of 0.1–2.0 m. The system contains four resonant tanks, which isolate from each other to maintain a high Q factor and stable, resonant angular frequency ω_0. The large air gap and small inductance L_m between the TX and RX tanks in an MRC system also enable a high Q factor and stable angular frequency to be maintained. However, this system requires considerable space for the installation of a power pad without ferrite and exhibits a strong leakage magnetic field.

CHAPTER 6

EV Charging Technology

Abstract: Wireless power transfer (WPT) systems for electric vehicles (EVs) are convenient and safe. In these systems, the inverter generates a high-frequency current and strong magnetic field, which are transferred to a primary compensated network. Simultaneously, the load absorbs power through a pickup compensated network, rectifier, and filter. Two types of WPT charging systems exist: static and dynamic. A static battery charging system stores considerable power during battery charging. This vehicle can be traveled over a long distance by using a heavy electric storage system. For example, the Tesla Model-S electric storage system, which weighs approximately 900 kg, can transfer power of up to 85 kWh, resulting in a driving range of 400 miles. However, a heavy electric storage system has low efficiency and causes pollution. A dynamic battery charging system has a smaller battery, lower weight, and higher efficiency than a static charging system. However, dynamic battery charging systems must be supported by integrated infrastructure, such as numerous grounded chargers. The source of power for static and dynamic charging is the power grid. Grid power mainly originates from centralized power factories, which generate power and emissions through the burning of fossil fuels. Renewable energy sources for powering EVs include sunlight, wind, rain droplets, static electricity, free lifting force, and regenerative brakes during driving. These renewable power source can be developed over the next decade years.

Keywords: Dynamic battery charging system, Emissions, Geenhouse effect, Regenerative brake, Renewable power sources, Static battery charging system.

1.1. INTRODUCTION

A typical electric vehicle (EV) is displayed in Fig. (**6.1**). An EV contains a plug-in battery charging system and wireless battery charging system, which adhere to the SAE J1772 [87 - 97] and J2954 [101, 221] standards, respectively. The aforemen tioned systems absorb energy through the same grid and power factor correction (PFC) unit. The power grid provides stable and cheap power [222 - 224] for charging the batteries of EVs. The PFC unit rectifies and tunes the absorbed power from the power source, which causes the power factor to be >0.85 and to be efficient [225, 227]. After PFC, the plug-in battery charging system outputs DC power to the D/D of the EV through a charging pile.

Chih-Cheng Huang & Chun-Liang Lin

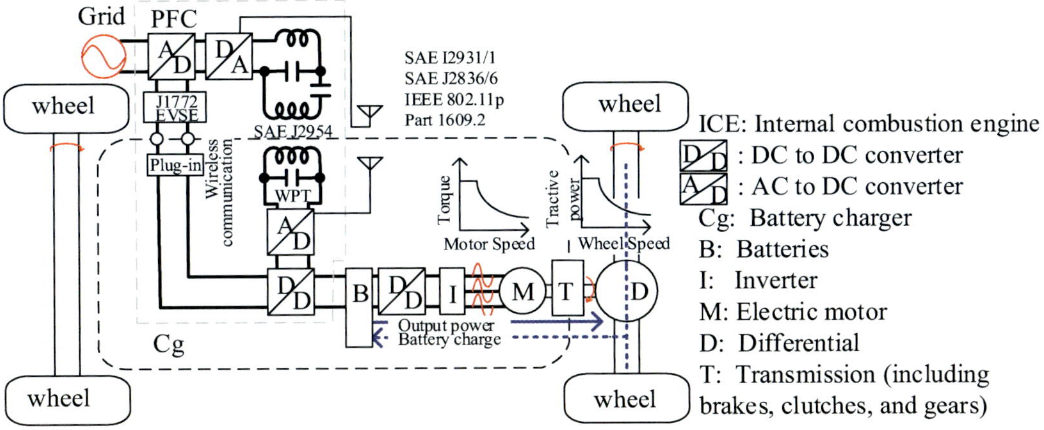

Fig. (6.1). Block diagram of a typical wireless charging system for EVs.

In wireless battery charging, a Direct Current to Alternating Current (D/A) unit generates a high-frequency current, which is transferred to an inductor–capacitor (LC) tank (compensated network). The LC tank amplifies the current or voltage to generate a strong magnetic field to link the LC tank of the pickup-side. The pickup-side battery absorbs power from the LC tank, a rectifier unit, and a DC to DC converter (D/D) unit. The compensated network amplifies the current or voltage with a high quality factor (*Q* factor) to generate a strong magnetic field [185, 201 - 203]. The D/D unit regulates the current and voltage in accordance with the battery type [228, 229]. The wireless battery charger system controls the charging voltage and current through communication with a radiofrequency (RF) device [129, 230] or novel data attach to operating frequency method [231, 232]. However, The RF device is complex and costly.

The International Commission on Non-Ionizing Radiation Protection (ICNIRP) 2010 guidelines describe the exposure safety boundaries for a wireless power transfer (WPT) system [102, 233].

Various standards have been published regarding the wireless charging of EVs. For example, the SAE J2954/1 charging standard for light-duty vehicles defines four charging levels in accordance with the power levels described in [101, 221, 234]. Table **6.1** summarizes the requirements of the J2954 standard for WPT systems.

Table 6.1. Proposed WPT Classes in the J2954/1 Standard.

Item	Phase	Voltage [V]	Max. Apparent Power [kVA]	Min. Target Efficiency [%]	Operating Freq. [kHz]
WPT1	signal	120	3.7	>85	81.38-90
WPT2	signal	240	7.7	>85	81.38-90
WPT3	signal	240	11.1	>85	81.38-90
WPT4	three	240	22	TBD	81.38-90

1.2. WPT with Grid

Fig. (**6.2**) displays a static battery charging system installed in a parking lot. In this figure, an EV enters the parking lot, and a giant magnetoresistance (GMR) system aligns its receiver coil with a transmission coil [236]. An EV can also be charged using a dynamic battery charging system during its movement on a road (Fig. **6.3**). A dynamic battery charging system contains a small battery, has low weight, and is highly efficient.

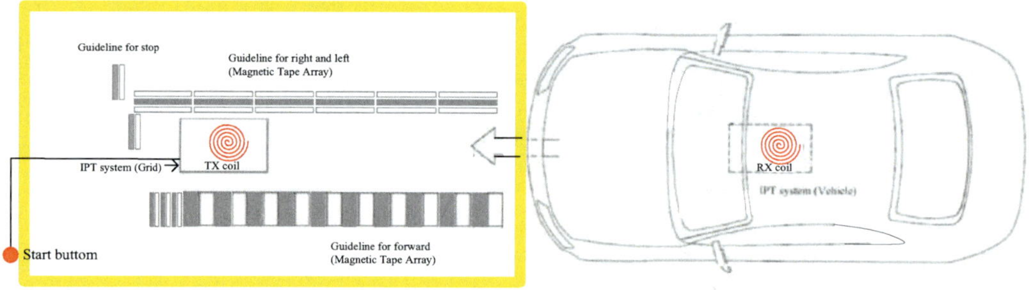

Fig. (6.2). Wireless power charging-discharging scheme combined with the GMR guidance system.

In a dynamic battery charging system (Fig. **6.3**), the car is transferred from charger 1 to charger 2 and then from charger 2 to charger 3. Such a system must be supported by integrated infrastructure, such as numerous grounded chargers [236, 237]. This requirement for integrated infrastructure results in high costs and limits the development of dynamic EV charging. Moreover, dynamic charging increases Electromagnetic Interference (EMI) [205, 238]. Power grids powered by centralized power factories are the primary power support for plug-in and wireless battery charging. These factories generate power by burning fossil fuels, which results in emissions and the greenhouse effect.

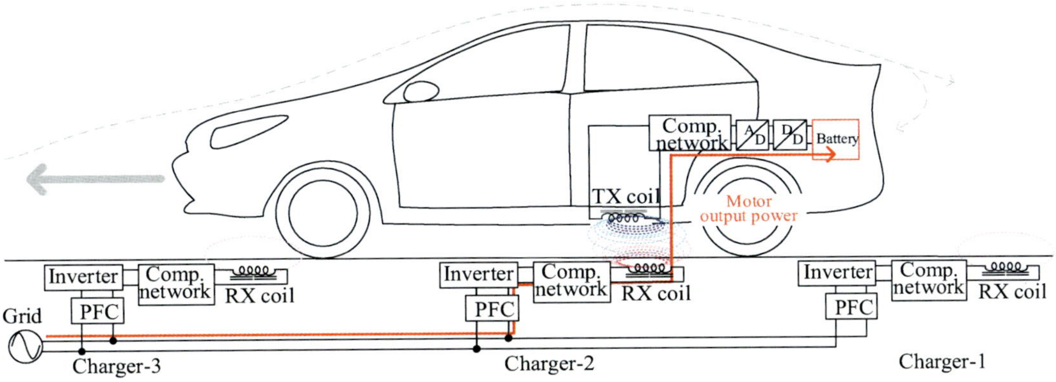

Fig. (6.3). A moving EV with dynamic charging system.

1.3. WPT with a Renewable Resource

Fig. (**6.4**) illustrates a family EV with a dynamic charging system and that is driven on a flat surface at a constant speed of 90 km/h. The power required by this EV from a motor for the aforementioned movement is calculated as follows:

$$P = \frac{Force_T \cdot v}{\eta} \quad \textbf{(6.1)}$$

where P is the motor power (W); $Force_T$ is the total force acting on the EV (N), including the rolling resistance, gradient resistance, and aerodynamic drag resistance; v is the velocity of the car (m/s); and η is the overall transmission efficiency, which is usually 0.7–0.9. The power required by a family EV moving on a flat surface with a constant speed of 90 km/h (Fig. **6.4**) under an aerodynamic resistance force of 200 N, rolling resistance force of 440 N, and overall transmission efficiency of 0.8 (gear loss) can be calculated as follows:

$$\left(90 \cdot \frac{km}{h}\right) \times \left(1000 \cdot \frac{m}{km}\right) \times \left(\frac{1}{3600} \cdot \frac{h}{s}\right) \times \frac{1}{0.8} = 15\text{kW} \quad \textbf{(6.2)}$$

The EV displayed in Fig. (**6.5**) is powered by renewable energy sources, such as sunlight, wind, rain droplets, static electricity, free lifting force, and regenerative brakes. These sources reduce the dependence of an EV on fossil fuels.

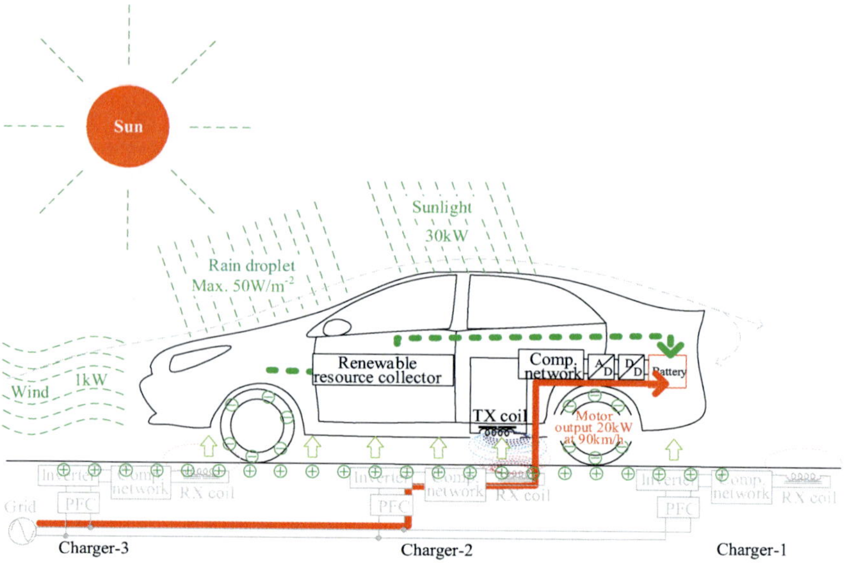

Fig. (6.4). Moving EV with a dynamic charging battery by renewable energy.

The EV receives solar irradiation of up to 2300 Wh/m^2 [239]. Solar cells using conventional III–V semiconductor compounds as photovoltaic materials absorb sunlight and convert it into electric power [240, 241]. Although some experimental solar cells have efficiency close to 50%, the efficiency of most commercial solar cells is lower than 30% [242]. Solar cells with efficiency of 20% that cover an area of 1.5 m^2 on an EV can generate a power of 30 kW.

Wind turbines convert wind power into electric power [139, 243 - 244]. Equations (6.2) and (6.3) describe the conversion of wind power into electric power.

$$P_{avail} = \frac{1}{2}\rho A v^3 C_p \quad \textbf{(6.3)}$$

$$A = \pi r^2 \quad \textbf{(6.4)}$$

where ρ is the air density, *A* is the sweep area of the turbine's blade, *v* is the wind speed, and C_p is the power coefficient. For example, with a blade length of 0.3 m, sweep area of 0.942 m^2, wind speed (*v*) of 13 m/s, air density of 1.23kg/m^3, and power coefficient (C_p) of 0.4, the energy generated by two micro wind turbines installed on the windward side of an EV can be expressed as follows:

$$P = \frac{1}{2} \times 1.23 \times 0.942 \times 12^3 \times 0.4 \times 2 = 1\text{kW}$$

In a droplet-based electricity generator [245, 246], water droplets continually impinge on a fluorinated material, which induces high charge density on the material's surface (Fig. **6.5**). A droplet-based electricity generator is fabricated by drop-casting polytetrafluoroethylene with a small quantity of aluminum onto a glass substrate coated with indium tin oxide. A study obtained an instantaneous peak power density of 50.1 W/m after the continual impinging of approximately 1.6×10^4 droplets.

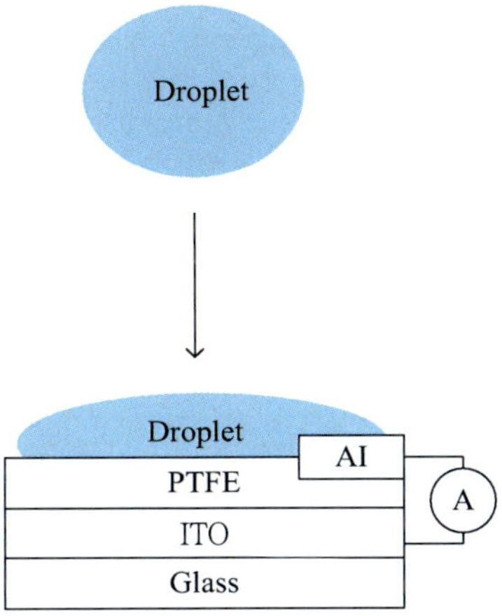

Fig. (6.5). Schematic of a droplet-based electricity generator (PTFE: polytetrafluoroethylene, ITO: indium tin oxide).

A regenerative braking system generates electric power through the movement of the EV containing it [247, 248]. The movement of an EV generates a free lifting force, and the contact of wheels with the ground generates static electricity. Cheap and stable power can be generated for EVs from the free lifting force and static electricity. However, this power cannot fully meet the requirements of a moving EV. Solar and wind energy can fully power a family EV that moves at 90 km/h on a flat roadway when the EV does not have a dynamic charging system.

CONCLUSION

Renewable power sources for powering moving EVs include sunlight, wind, rain droplets, static electricity, free lifting force, and regenerative brakes, which can be developed over the next decade years. This chapter describes the feasibility of powering EVs by using renewable sources. A family EV that must be driven on a flat surface at a constant speed of 90 km/h requires a dynamic charging system that can generate a power of 15 kW. Solar cells and micro wind turbines can generate 30 and 1 kW of energy, respectively. This power can be generated without support from a power grid.

CHAPTER 7

Adaptive Compensation of Magnetic Inductive Variation

Abstract: Because of increased awareness regarding environmental protection and decreased cost of electric vehicle (EV) batteries, the market share of EVs has gradually increased. EVs can cogenerate power with a traditional centralized power generation system. A bidirectional wireless power transfer battery charging system (BWPTBCS) in an EV is a distributed energy storage system that acts as a virtual power plant. This virtual power plant instantly complies with the command of an aggregator to cogenerate energy with the traditional centralized power generation system during peak demand hours. Because energy storage systems respond quickly to the load, the spinning reserve of the traditional centralized power generation system is reduced. Moreover, energy storage systems distribute power over a city and near the load, which can reduce the energy transmission distance from the traditional centralized power generation system and increase the energy transmission efficiency. Therefore, an EV with a BWPTBCS is beneficial for consumers, the traditional centralized power generation system, and owners. This chapter proposes an adaptive-compensation-based magnetic induction system (ACMIS) for BWPTBCSs. The proposed ACMIS is based on zero-voltage switching, transmits power from a single-end inverter of a Parallel–Parallel (PP) type compensated network, and automatically adjusts the impedance of an inductor–capacitor tank to that of a matching network. The ACMIS efficiently transmits power wirelessly under a varying air gap and in the case of misalignment between two power pads.

Keywords: Adaptive-compensation-based magnetic induction system (ACMIS), Aggregator, Bidirectional wireless power transfer battery charging system (BWPTBCS), Distributed energy storage system, Impedance matching network, Single-end inverter, Spinning reserve, Virtual power plant, Zero-voltage switching (ZVS).

7.1. INTRODUCTION

Electric vehicles (EVs) act as virtual power plants that can support traditional power plants during peak demand hours. These vehicles are beneficial to consumers, traditional centralized power plants, and owners.

An EV with BWPTBCS drives on the road with a wireless power change system's infrastructure, as shown in Fig. (**7.1**). In the figure, the red line illustrates power

Chih-Cheng Huang & Chun-Liang Lin

flow from grid to vehicle. This EV collects nearby renewable sources such as sun, wind, *etc*., to store energy in a battery or provide feedback to the grid as a green line during peak domain hours.

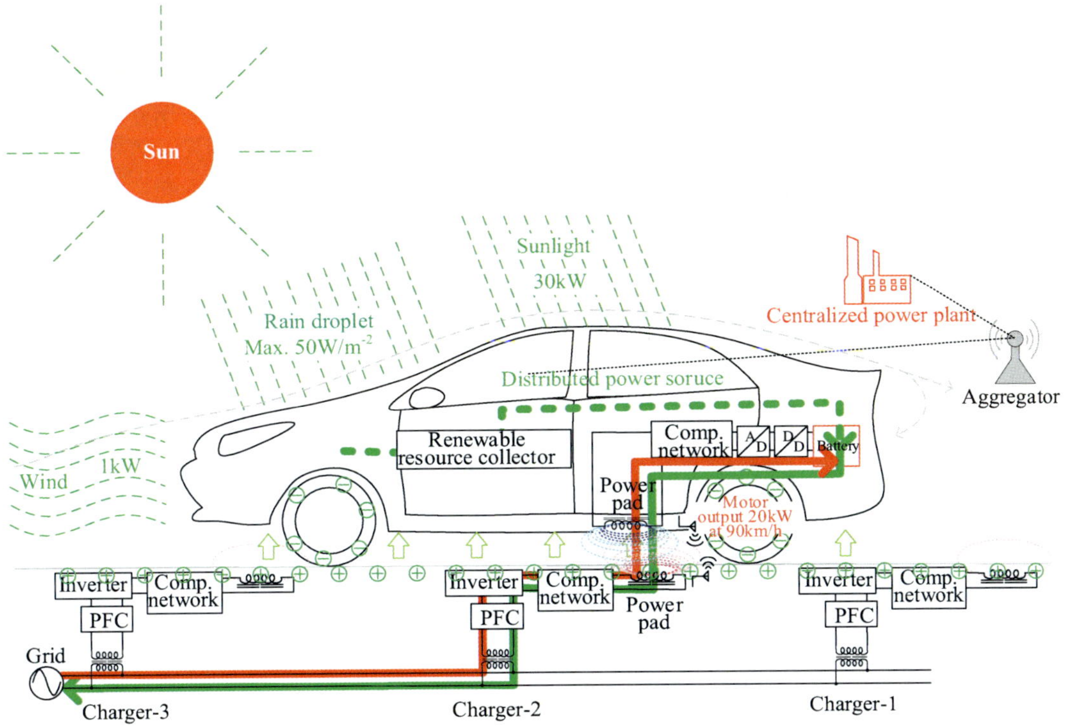

Fig. (7.1). EV with a bidirectional wireless power transfer battery charging system that is moving on a road and powered by renewable energy.

The aggregator collects them to make a virtual power plant, which helps traditional power plants in generating electricity more efficiently and reducing emissions. However, wireless power transfer on two pads was affected by varying air gaps and misalignment, which decay power transfer and efficiency. This chapter suggests that ACMIS can efficiently keep a large amount of power transfer to immunize varying air gaps and misalignment between the two power pads. Besides, ACMISs can instantly respond to loads that aid traditional power plants to reduce the power grid's spinning reserve.

7.2. Bidirectional Power Flow Configuration on an EV

Fig. (**7.2**) depicts a BWPTBCS with the proposed ACMIS. The ACMIS contains a primary unit and a secondary unit. The primary unit's power pad is installed underground below a parking lot or road, and the secondary unit's power pad is installed in the chassis of the EV. The battery charging system has an efficiency of 0.8, a power rate of 1.5 kW, and a battery capacity of 30 kWh. An EV with the aforementioned charging system can drive 180 km, managing to drive 6 km for 1 kWh of power. This EV can be driven for 90 km when its battery is charged for 10 h.

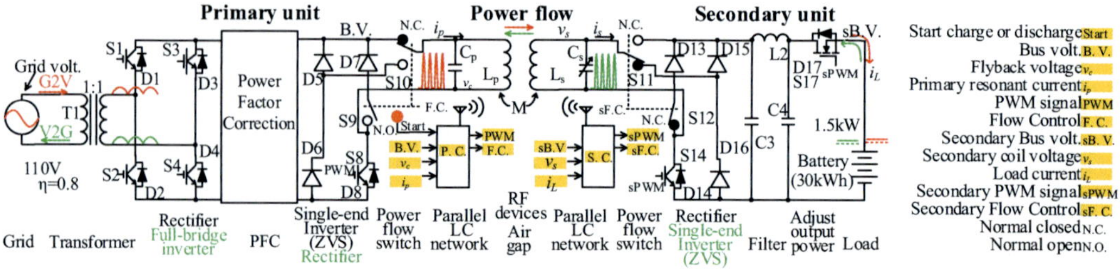

Fig. (7.2). BWPTBCS with the proposed ACMIS.

The proposed ACMIS can drive grid-to-vehicle (G2V) or vehicle-to-grid (V2G) power flow. In the system, a favorable power factor is achieved through power factor correction (PFC). A single-end inverter, which is mainly composed of an insulated-gate bipolar transistor, generates a 15–30-kHz current through zero-voltage switching (ZVS) [110, 249, 250]. This inverter is smaller and cheaper than a full-bridge inverter. In a single-end inverter, a switch is turned on when the fly-back voltage is 0 V so that the charger has high efficiency and a small cooling fin.

To generate high fly-back voltage and high current in a single-end inverter, its parallel inductor–capacitor (LC) compensated network has a Q factor higher than 2. A magnetic link is used to achieve high power transfer to the load.

Power flow switches dictate the direction of power flow, and a filter is used to reduce switch noise and the intensity of high-order current harmonics. The output power is adjusted by varying the charging speed using a MOSFET with a switching frequency of approximately 100 kHz. The radiofrequency transceivers communicate between the primary controller (P. C.) and secondary controller (S. C.). This kind of communication includes power flow direction, charging voltage data, charging current data, and charging watt data.

When the charging system conducts G2V power transmission, the power flow path is as follows: grid → T1 → rectifier (D1–D4) → PFC device → single-end inverter (S8) → power flow switch (S9's state is normally open (N. O.) and S10's state is normally closed (N. C.) → parallel LC network (C_p, L_p) → air gap → parallel LC network (C_s, L_s) → power flow switch (the states of S11 and S12 are normal-closed) → rectifier (D13–D16) → filter → S17 → load (battery).

When the charging system conducts V2G power transmission, the power flow path is as follows: battery → D17 → filter → single-end inverter (S14) → power flow switch (S11's state is changed to closed and S12's state is changed to open) → parallel LC network (C_s, L_s) → air gap → parallel LC network (C_p, L_p) → rectifier (D5–D8) → PFC device → full-bridge inverter (S1–S4) → T1 → grid.

7.3. Adaptive Tracking of the Optimal Switching Frequency Using an Impedance Matching Network

A p–p-type compensated network can generate a strong current and magnetic field for efficiently transmitting high power wirelessly. However, wireless power flow decays under a varying air gap and power pad misalignment because the mutual inductance (M) reflects the impedance of the load, including real and imaginary parts. The proposed ACMIS exhibits the advantage of a p–p-type compensated network and overcomes the power flow decay problem.

Fig. (**7.3a**) illustrates G2V power transmission with the ACMIS. If the size and inductance (L_P) of the primary unit's power pad are determined, the compensated $C_P$can be calculated as follows [112, 232]:

$$C_p = \frac{(L_p L_S - M^2) C_S L_S^3}{M^4 C_S R_L + L_S (L_p L_S - M^2)^2} \tag{7.1}$$

The inductance and capacitance of the secondary unit's power pad [L_s and C_s (C_{s-1}), respectively] are equal to L_p and C_p, respectively. An impedance matching network [251, 252] is composed of C_{s-1}, C_{s-2}, and S18. These symbols, including C_{s-1} and $C_{s-2,}$ are displayed in Fig. (**7.3 b**). In this figure, the state of S18 is normal-open, and $C_s = C_{s-1}$. When the state of S18 is changed from open to closed, $C_s = C_{s-1} + C_{s-2}$.

The voltage and current curves of the compensated network are shown in Fig. (**7.4**). This figure illustrates four cycles of T. For example, in the first cycle, S8 is turned on by a rising-edge signal at t_0, high-frequency current is transferred to the LC compensated network, and i_p (primary LC tank current), V_C (primary LC tank voltage), i_s, and V_s (secondary LC tank voltage) are generated. The parameter i_p can be increased or decreased by increasing or decreasing the pulse width of the pulse width modulation signal, respectively, to transfer energy to the inductor with an inductance of L_p and the capacitor with a capacitance of C_p. When the pulse signal is turned off at t_1, L_p and C_p generate fly-back voltage and a strong magnetic field to link the secondary unit's LC compensated network with mutual inductance and magnetic field between t_1 and t_2 (t_{fb}). The parameter t_{fb} is affected by the mutual inductance and load. An example of a load is a lithium-ion battery, the charging curve of which [253] is presented in Fig. (**7.5**).

Fig. (**7.5**) Illustrates the following two main stages: an initial constant-current (CC) mode and a subsequent constant-voltage (CV) mode. The charging process begins from the CC mode, and the battery voltage increases rapidly to a specified level. Then, the charger enters the CV mode until the charging current decreases to nearly 0 Amp. Therefore, t_{fb} is not fixed and causes variation in the charging curve.

The next rising-edge pulse signal turns on S8 by decreasing t_{fb} at a V_C value less than the bus voltage. This process is called ZVS. The signal-end inverter, which contains a parallel LC compensated network, always takes varying mutual inductance and reflect impedance each *T* period when using the ZVS method. The proposed ACMIS efficiently transmits considerable power wirelessly.

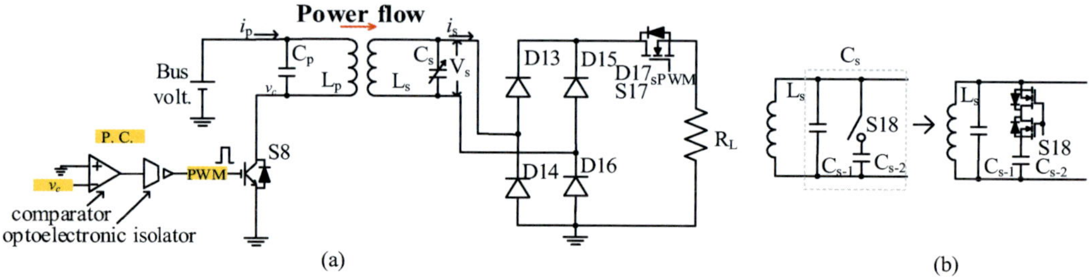

Fig. (7.3). (**a**) Configuration of the proposed ACMIS and (**b**) topology of an impedance matching network.

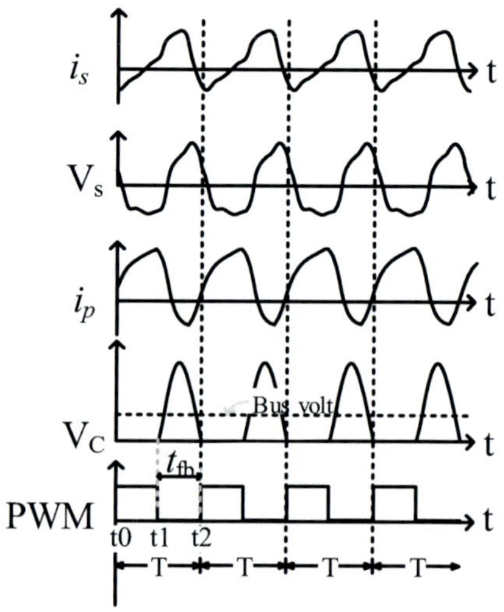

Fig. (7.4). Critical voltage and current curves.

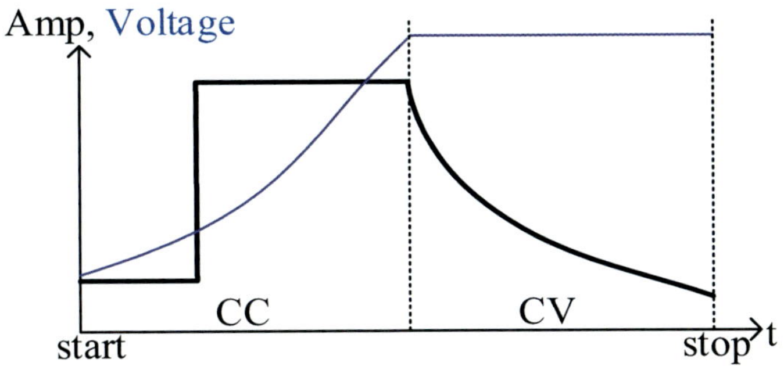

Fig. (7.5). Typical charging profile of a lithium-ion battery cell.

Fig. (**7.6**) displays voltage gain versus operation frequency curves for the proposed ACMIS. As depicted in this figure, the natural resonant frequency of a p–p-type compensated network (f_0) is approximately 23 kHz. The figure contains curves for k = 0.15, 0.20, 0.25, 0.55, 0.60, and 0.70. When k = 0.15–0.25, the voltage gain gradually increases as the air gap between the two power pads decreases. When k = 0.55–0.70, the voltage gain distinctly decreases as the air gap between the power pads decreases. The six red dots close to the peak point (optimal point) are the operating points of the ACMIS for ZVS. However, two curves, namely those for k = 0.60 and 0.70, represent frequency bifurcation [112, 211, 254]. The operating points of the ACMIS on these curves are far from the optimal point. To overcome this problem, the controller in the primary unit is used to turn on S18 in the impedance-matching network to increase C_s when k = 0.6–0.7. This step causes an increase in the secondary side's natural resonant frequency, a reduction in frequency bifurcation, and a rearrangement of the ACMIS's optimal operating point.

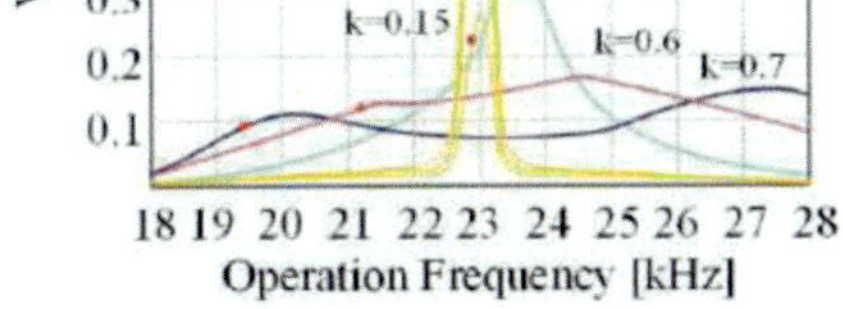

Fig. (7.6). Voltage gain versus operation frequency for the ACMIS.

CONCLUSION

This chapter proposes a BWPTBCS with the proposed ACMIS, which can transmit energy between EV and power grid. The proposed ACMIS is a compact charger with operation based on the ZVS method. The system automatically compensates for the impedance-matching network of an LC tank to overcome the problems caused by a varying air gap and misalignment between two power pads.

CHAPTER 8

Coherent Wireless Power and Data Transfer

Abstract: A wireless electric vehicle (EV) battery charger contains two units: a primary unit and a secondary unit. The charger conducts grid-to-vehicle or vehicle-t--grid power transmission by using a magnetic link. In general, the two units of the aforementioned charger communicate through radiofrequency communication technology, such as Bluetooth or Wi-Fi. The data transferred in such communication include the EV's identification number, battery status, and active control commands. The use of communication devices results in increased costs. Moreover, radiofrequency communication is associated with signal interference and low real-time performance. This chapter proposes a method for coherent wireless power and data transfer. The method includes handshaking communication, current trimming, and the attachment of data to modulated digital data by using a magnetic link. This method also supports the transmission of real-time messages from the secondary unit through a carry wave.

Keywords: Bluetooth, Current trimming method, Data attachment method, Grid to vehicle (G2V), Handshaking communication, Vehicle to grid (V2G), Wi-Fi.

8.1. INTRODUCTION

A wireless electric vehicle (EV) battery charger conducts vehicle-to-grid (V2G) or grid-to-vehicle (G2V) power transmission in parking lots or roads, respectively. To achieve such transmission, the charger's primary and secondary units should be magnetically linked. Wireless EV chargers are safe and efficient and reduce a power unit's emissions during charging and discharging. This chapter proposes a method for coherent wireless power and data transfer (Fig. **8.1**). The method includes simultaneous power and digital data transfer through handshaking communication [255], current trimming [232], and data attachment [231] without the use of radiofrequency communication devices.

Before power transfer, the primary unit receives data such as the EV's identification number and battery status through handshaking communication. The system begins charging or discharging the battery after handshaking between the primary and secondary units has been confirmed.

Chih-Cheng Huang & Chun-Liang Lin

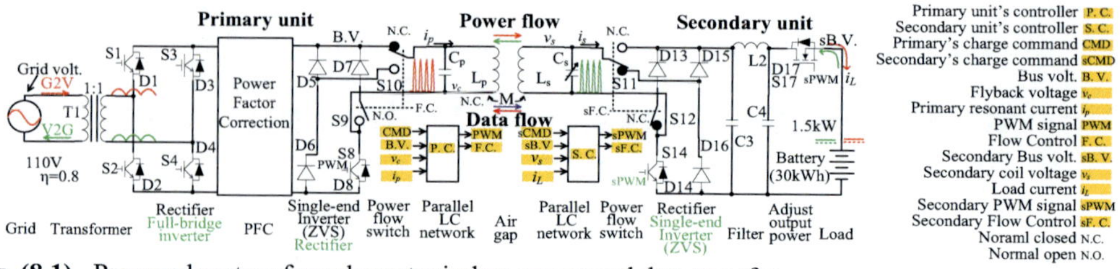

Fig. (8.1). Proposed system for coherent wireless power and data transfer.

After the power transfer, two channels exist for data transmission, a general channel and an emergency channel. In the general channel, the primary unit commands the secondary unit with a low data transfer rate by using the current trimming method to modulate digital data transfer. The secondary unit feeds back data such as the charging voltage, charging current, and charging power, *etc.*, to the primary unit by using a novel data attachment method. If an emergency such as overvoltage or overcurrent occurs in the secondary unit, the unit sends a signal through the emergency channel to the primary unit instructing it to quickly stop power transfer.

A driver can employ a parking guidance system to align two power pads for efficiently charging an EV battery in a parking lot [235].

8.2. Bidirectional Power Transfer

The system illustrated in Fig. (**8.1**) controls bidirectional power flow through switches, as shown in Fig. (**8.2**). Table **8.1** lists the statuses of the switches displayed in Fig. (**8.2**) in various states. A schematic representation of the system in the G2V state is displayed in Fig. (**8.2 a**). In this state, the inverter (S8) on the primary side functions in accordance with the zero-voltage switching (ZVS) method [110, 249, 250] and generates a high-frequency current that is transferred to the primary side's parallel inductor–capacitor (LC) network. The secondary side's parallel LC network then receives this current through magnetic induction. Subsequently, the aforementioned current is transferred to a battery (load) through rectifiers (D13–D16), a filter, and an output power switch (S17).

Table 8.1. Statuses of the switches of the proposed system in different states.

Two Stage		Source	Load	S1~S4	S8	S9	S10	S11	S12	S14	S17
Before Power Transfer	Handshaking G2V	-	-	Rectifier	Inverter	N.O.	N.C.	N.C.	N.C.	Diode	Converter
	Handshaking V2G	-	-	Rectifier	Diode	Turn-closed.	Turn-on	Turn-on	Turn-open	Inverter	Diode
After Power Transfer	G2V	Grid	Batteries	Rectifier	Inverter	N.O.	N.C.	N.C.	N.C.	Diode	Converter
	V2G	Batteries	Grid	Inverter	Diode	Turn-closed	Turn-on	Turn-on	Turn-open	Inverter	Diode

A schematic representation of the system in the V2G state is displayed in Fig. (**8.2 b**). In this state, the inverter (S14) on the secondary side functions in accordance with the ZVS method. It generates a high-frequency current that is transferred to the secondary unit's parallel LC network. The primary unit's parallel LC network then receives this current through magnetic induction. Subsequently, the aforementioned current is transferred to the power grid through rectifiers (D5–D8), a power factor correction device, inverters (S1, S2, S3, and S5), and an isolated transformer.

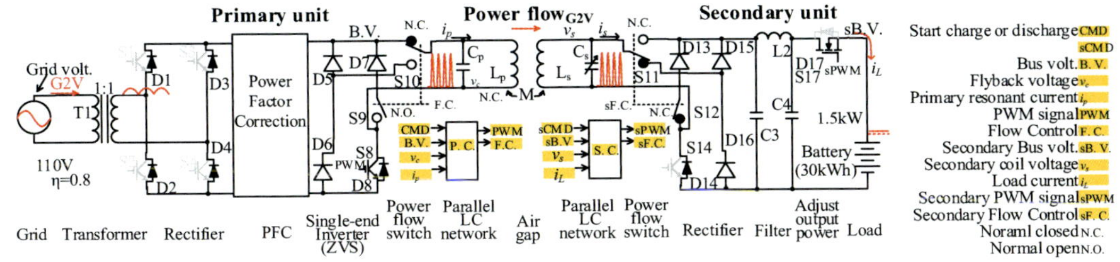

(**a**) Schematic representation of the proposed system in the G2V state.

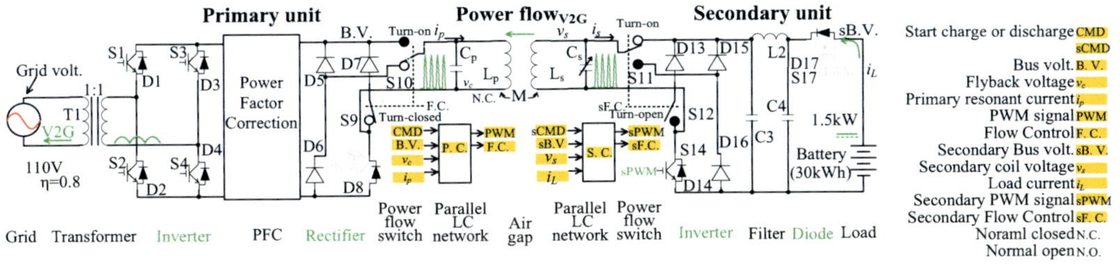

(**b**) Schematic representation of the proposed system in the V2G state.

Fig. (8.2). Bidirectional power transfer in the proposed system.

8.3. Data Transfer

8.3.1. Modulation Principles

Two primary types of data modulation are introduced in the following text: amplitude modulation (AM) and frequency-shift keying (FSK) modulation. The block diagrams for these two types of data modulation are illustrated in Fig. (**8.3**).

In AM (Fig. **8.3 a**), two signals, namely v_1 and v_2, are input to an amplitude modulator. The v_1 signal is a serial signal whose data bits in sequence are 0101011. The v_2 signal is a carry wave that carries the signal and transmits power to the target through the inverter and LC tank 1. The load absorbs energy from LC tank 2, a rectifier, and an amplitude demodulator. The amplitude demodulator conducts magnitude modulation. It separates the v_1 signal from the carrier wave (v_2). The signals v_1 and v_1' are the same. However, v_1' exhibits a phase shift, thus, a magnitude distortion occurs compared with v_1. This magnitude distortion may affect the received correction of data.

The configuration for FSK modulation (Fig. **8.3 b**) is similar to that for AM; however, a modulator and demodulator are used in FSK modulation. In FSK modulation, the carry wave frequency for the bit "0" is less than that for the bit "1." The parameter v_1' in Fig. (**8.3 b**) exhibits a phase shift and magnitude distortion compared with v_1. However, the magnitude's distortion does not affect the data value because FSK modulation involves frequency modulation and not magnitude modulation.

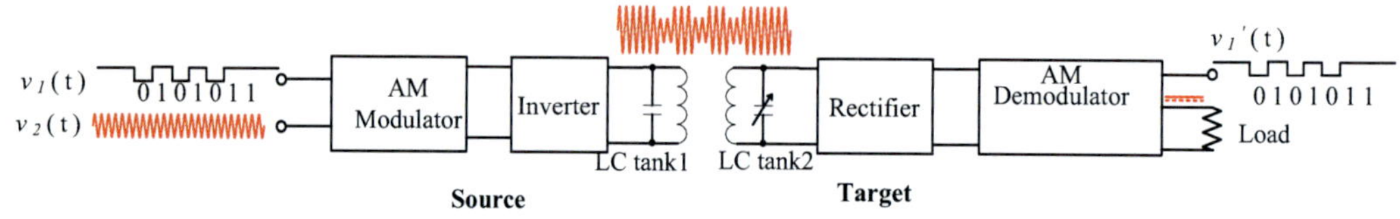

(**a**) Wireless data communication through AM modulation

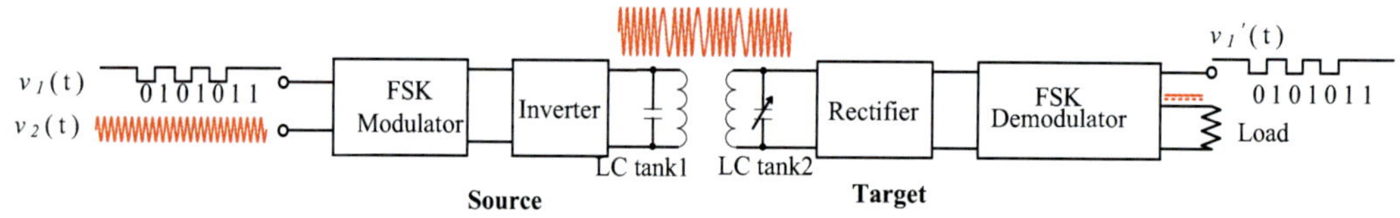

(**b**) Wireless data communication through FSK modulation

Fig. (8.3). Two types of data modulation.

Fig. (**8.4**) displays the power and data network between the EV, aggregator, primary and secondary units, and power grid. Two communication levels exist in the EV, outer and inner communication. In outer communication, a cell phone receives commands and translates them. The cell phone is linked to a charger through a primary side's command input port (CMD) and the secondary side's command input port (sCMD) (Fig. **8.2**). Inner communication is divided into two stages, the inner communication before and after the power transfer stage. Before the power transfer stage, the proposed system conducts handshaking communication to modulate digital data. The primary unit requires information on the battery status, and the secondary unit requires information on aspects such as the battery capacity. On the basis of the battery status, the primary unit conducts G2V or V2G power transmission. After the power transfer stage, the proposed system uses the current trimming and data attachment methods to modulate digital data. In these methods, the general data channel is used to transmit data, such as the charging or discharging voltage, current, and power.

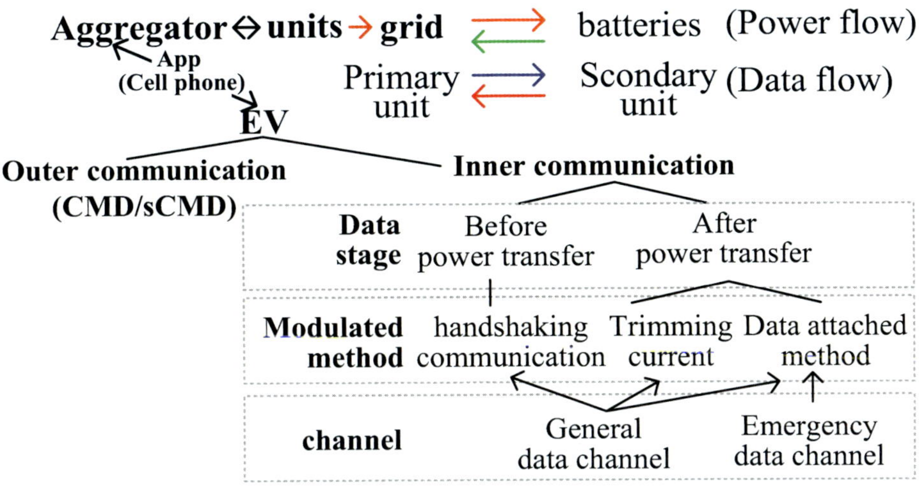

Fig. (8.4). Power and data network.

8.4 . Before Power Transfer

Before power transfer, the charger decides the type of power transmission—whether G2V or V2G power transmission—through handshaking inner communication. A handshaking data packet, such as that displayed in Fig. (**8.5**), is transferred between the primary and secondary units. This packet comprises 16 bits and includes the command, checksum command, data, and checksum data packets. In Fig. (**8.5**), Checksum_{CMD} equals the command's complement + 1, and Checksum_{DAT} equals the data's complement + 1. The command and data are listed in Table **8.2**.

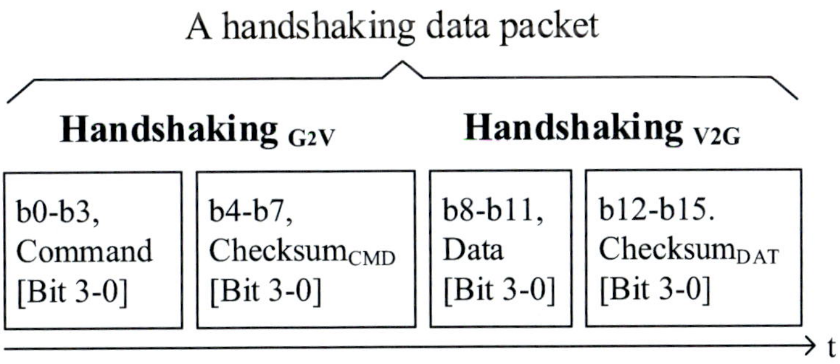

Fig. (8.5). Serial handshaking data packet.

Table 8.2. List of handshaking commands and data.

Handshaking$_{G2V}$ (Command) [b0, b1, b2, b3]	Handshaking$_{V2G}$ (Data) [b8, b9, b10, b11]	Description
0101	0111	The primary unit requires battery status, and the secondary unit acknowledges insufficient battery.
	0011	The primary unit requires battery status, and the secondary unit acknowledges sufficient battery.
	1111	The primary unit requires battery status, and the secondary unit acknowledges fault.
0010	0010	The primary unit commands power flow from G2V, and the secondary unit acknowledges pass.
	1111	The primary unit commands power flow from G2V, and the secondary unit acknowledges fault.
0011	1100	The primary unit commands power flow from G2V, and the secondary unit acknowledges pass.
	1111	The primary unit commands power flow from G2V, and the secondary unit acknowledges fault.

Handshaking communication is based on the principle of resonance [256] and is used to transmit and receive commands and data. The principle of resonance is illustrated in Fig. (**8.6**). In Fig. (**8.6 a**), the alternating current (i_1) decays slowly and produces a long free resonance time for the LC tank. In Fig. (**8.6 b**), i_1 is absorbed by the load (R_L), which results in a short free resonance time for the LC tank. The resonance time is used to detect the possible load and data transfer before conducting power transfer.

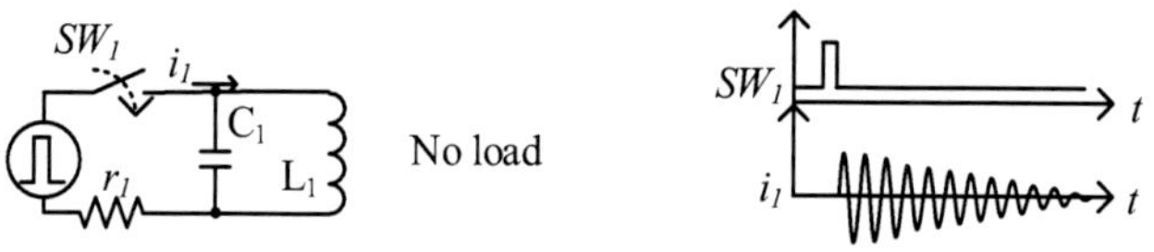

(**a**) Free resonant energy is transferred to an LC tank with a long resonance time

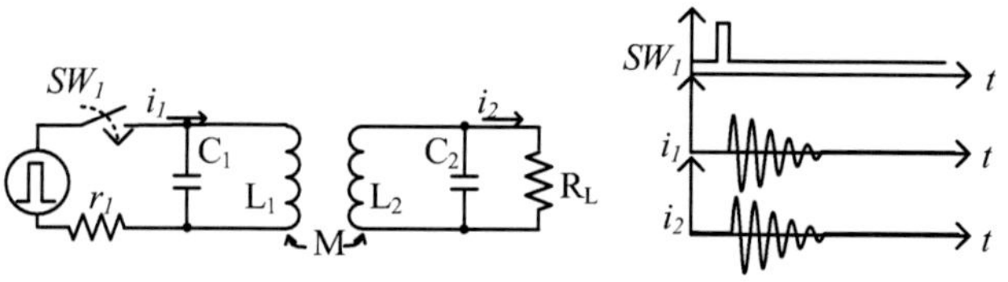

(**b**) Resonant energy absorbed by the load within a short resonance time

Fig. (8.6). Graphical explanation of the resonant principle.

The *handshaking communication circuit for G2V* power transmission is displayed in Fig. (**8.7**), and the corresponding timing diagram is illustrated in Fig. (**8.8**). In the aforementioned circuit, S8 is turned on to generate a pulse for producing the start bit "0." S8 is then turned off to generate the bit "1" during T for transmitting a command packet to the secondary unit through the LC tank by using parallel-parallel (PP) type architecture. The secondary unit receives the command packet through the Decode$_S$ pin. As displayed in Fig. (**8.8**), the primary unit sends the command packet "0b0101," and the secondary unit receives this command packet through the Decode$_S$ pin. The primary processor determines a suitable air gap by detecting $t_{free\_resonate}$ in the Decode$_P$ pin after receiving the start bit. For example, if a suitable air gap exists between the inductors with inductance L_P and L_S, $t_{free\_resonate} << T$ (= 16.66 ms) because the pulsed energy at the primary-side LC tank is absorbed by the load. If $t_{free\_resonate} > T$, energy remains within the primary-side LC tank, and the system halts power transfer.

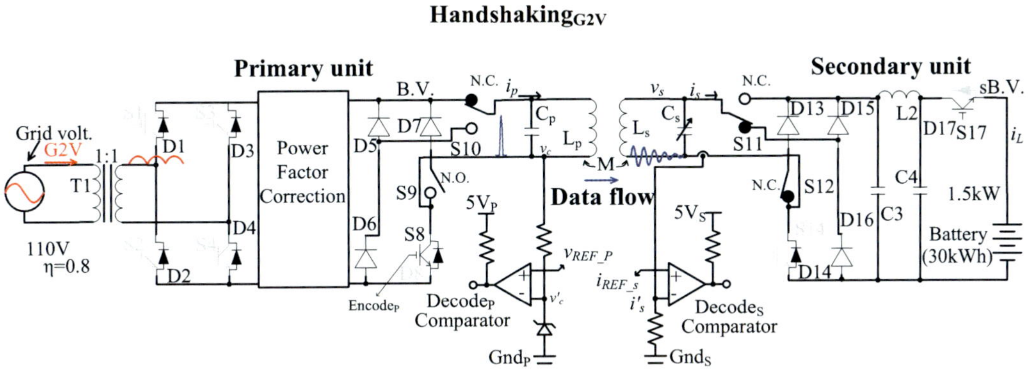

Fig. (8.7). Handshaking communication circuit for G2V power transmission.

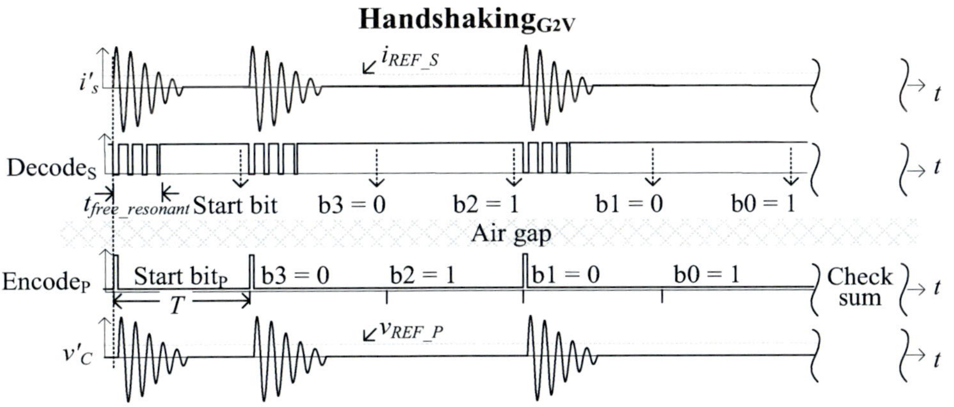

Fig. (8.8). Timing diagram of the handshaking communication circuit for G2V power transmission.

The handshaking communication circuit for V2G power transmission is presented in Fig. (**8.9**), and the corresponding timing diagram is displayed in Fig. (**8.10**). In the aforementioned circuit, S14 in the secondary unit is turned on to generate a pulse for producing the start bit "0." S14 is then turned off to generate the bit "1" during *T*. The primary unit receives digital bits through the Decode_P pin. As depicted in Fig. (**8.10**), the secondary unit sends the data packet "0b0111," and the primary unit receives this data packet through Decode_P. The primary unit then sends the command packet "0b0101" to acquire the battery status. Subsequently, the secondary unit responds by sending the packet "0b0111" to acknowledge insufficient battery capacity.

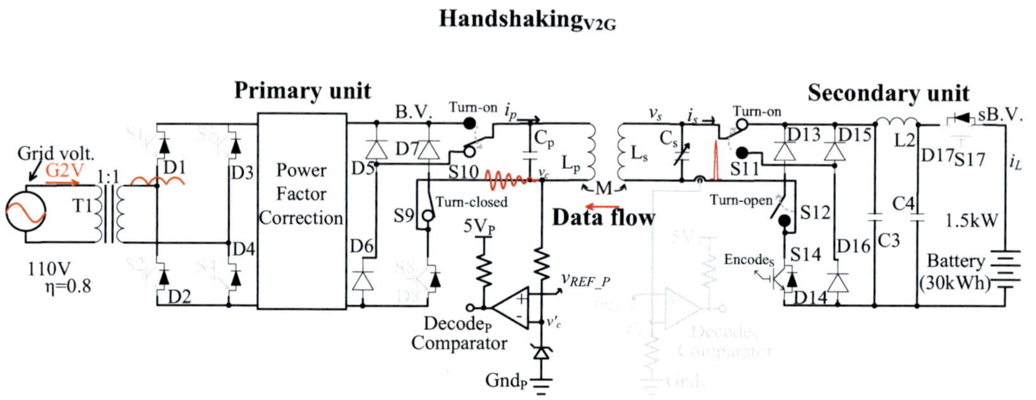

Fig. (8.9). Handshaking communication circuit for V2G power transmission.

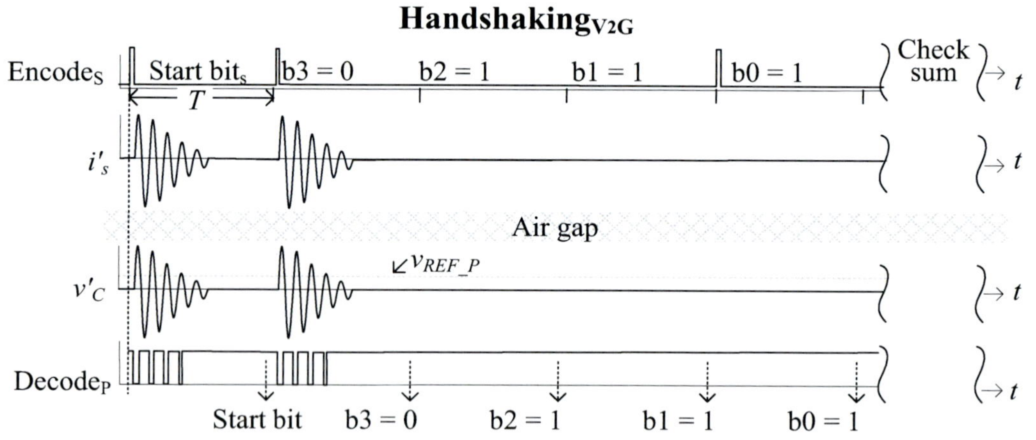

Fig. (8.10). Timing diagram of the handshaking communication circuit for V2G power transmission.

8.5. After Power Transfer

When charging or discharging the battery, power is carried by carry waves in G2V or V2G power transmission. This energy is combined with coherent wireless data transfer to achieve communication between the primary side and secondary side by using modulated data adopting the current trimming and data attachment methods. The modulated data are subjected to binary FSK (BFSK) modulation, in which a pair of discrete frequencies is employed to distinguish and transmit binary data, as displayed in Fig. (**8.11**).

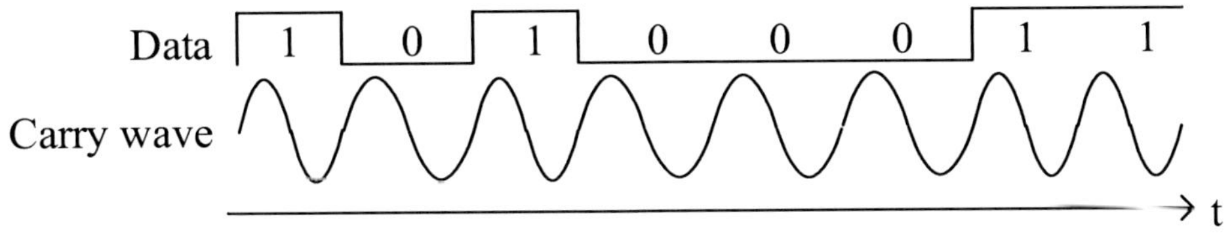

Fig. (8.11). Serial data flow in BFSK modulation.

These messages, including charging voltage, charging current, and charging power, *etc.*, are divided into two channels in accordance with its attributes. The first channel contains the general message, the EV's identification number, charging or discharging voltage, current, and power. The general channel is processed with a low data rate to decrease the signal interference caused by ZVS. The second channel contains the emergency message, which includes information regarding overvoltage or overcurrent status. This channel is processed with a high data rate to ensure a real-time reaction. Table **8.3** summarizes the specifications of the aforementioned two channels.

Table 8.3. Specifications of the general and emergency channels after power transfer in inner communication.

Channel	Baud Rate [Bd.]	Start Segment [bit]	Data Segment [bit]	Checksum Segment [bit]	Stop Segment [bit]
Normal	120	1	8	8	1
Emergency	Carry wave	1	4	4	1

To prevent data collision, the charger employs a requirement and acknowledgment mechanism for inner communication. The primary unit acquires the status of the secondary unit, and the secondary unit acknowledges this status. An example of a dataset transferred during power transmission is displayed in Fig. (**8.12**). The primary unit sends a transmission (Tx) command packet to the secondary unit to command the secondary unit to provide the battery's status. In accordance with this command, the secondary unit acknowledges the transferred data packets, which include information such as the EV's identification number and the battery's remaining capacity, output voltage, and current.

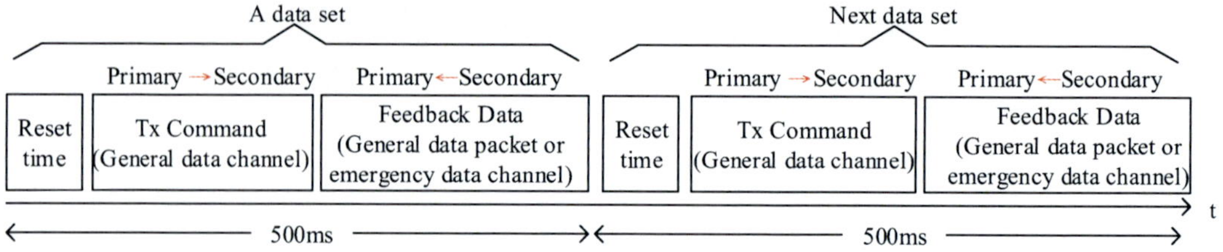

Fig. (8.12). Format of datasets transferred after power transmission in inner communication.

8.5.1. Current Trimming Method

After power transfer, the primary unit modulates commands by using the current trimming method, as illustrated in Fig. (**8.13**). The switch of S8 in this figure is not only influenced by ZVS [110] but also modulated data expressed in equation 8.1.

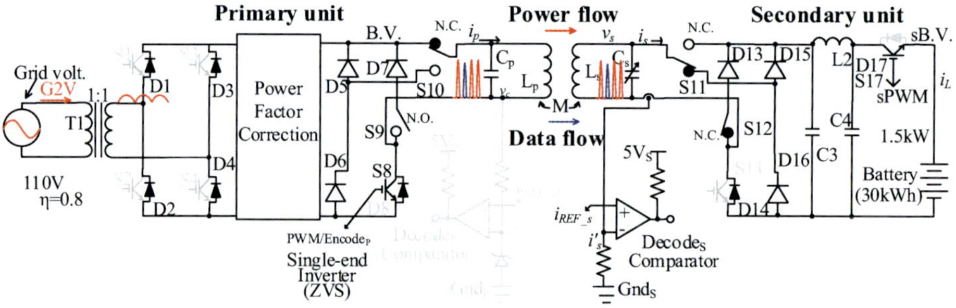

Fig. (8.13). Transceiver configuration in the current trimming method.

$$S(t) = \begin{cases} A\cos(2\pi f_{PWM} t), & "1" \\ A\cos\left(2\pi f_{trimming\ curve} t\right), & "0" \end{cases} \tag{8.1}$$

In the general data channel, the primary unit transmits a Tx command packet to the secondary unit. For example, in Fig. (**8.14**), the Tx command contains a start bit, b0, and b1. In this figure, S8, which is located in the primary unit, generates a pulse for simultaneously controlling the output current and modulating the bit "1" during *T*, as expressed in equation 8.2. At points A and B in Fig. (**8.14**), S8 outputs sorting pulses to modulate the start bit and b0, respectively. The shorter time compared with *T* is bit equal to '0', as expressed in equation 8.3, where Δτ is a default value.

$$S8 modulate/encode digit "1" = T \quad \textbf{(8.2)}$$

$$S8 modulate/encode digit "0" = T\text{-}\Delta\tau \quad \textbf{(8.3)}$$

The secondary unit uses Decode$_s$ to obtain T (also denoted t_{Bit}). The T value for the secondary unit is expressed in equations 8.4 and 8.5.

$$T = \frac{1}{4}\sum_{i=1}^{4} T_{Si}, \text{ during obtaining star bit} \quad \textbf{(8.4)}$$

$$T = \frac{1}{4}\sum_{i=1}^{4} T_{bit}, \text{ during obtaining data bits} \quad \textbf{(8.5)}$$

The secondary unit decodes the bits “1” and “0” as per equations 8.6 and 8.7, respectively. After finding the start bit, the secondary unit waits 4.16 ms and then searches the next bit every 8.33 ms. If t_{Bit} satisfies equation 8.6 in each 8.33 ms, this bit is “0”; otherwise, this bit is “1”.

$$\text{Secondary unit decodes digit "1"} = t_{Bit} \geq \left(T - \frac{\Delta\tau}{2}\right) \quad \textbf{(8.6)}$$

$$\text{Secondary unit decodes digit "0"} = t_{Bit} < \left(T - \frac{\Delta\tau}{2}\right) \quad \textbf{(8.7)}$$

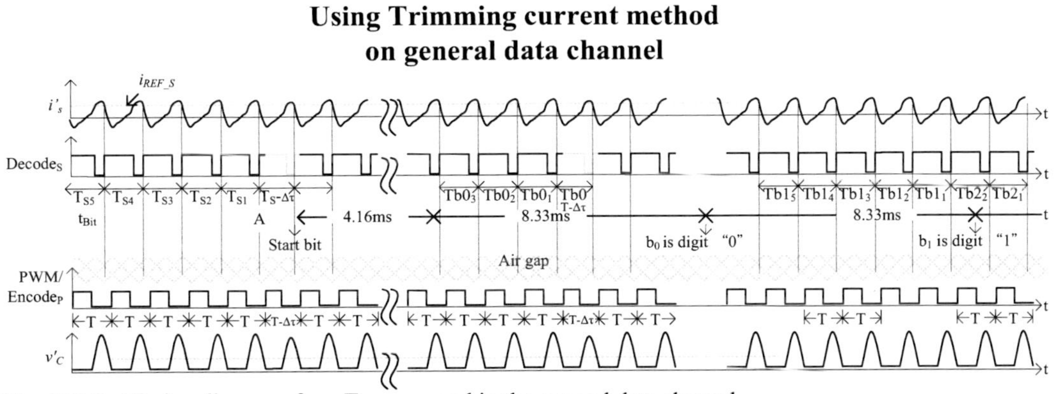

Fig. (8.14). Timing diagrams for a Tx command in the general data channel.

8.5.2. Data Attached Method

Following the TX command, the primary unit modulates feedback data in the general channel by using a novel data attachment method. The circuit used to implement this method is displayed in Fig. (**8.15**). The pulse width modulation (Encode$_s$) signals of S17 simultaneously control the output current and modulate the carry wave feedback. The format of the aforementioned data is expressed in equation 8.8.

$$S(t) = \begin{cases} A\cos(2\pi f_{sPWM}t), & \text{"1"} \\ A\cos(2\pi f_{data\ attech}t), & \text{"0"} \end{cases} \quad \textbf{(8.8)}$$

Novel data attached method

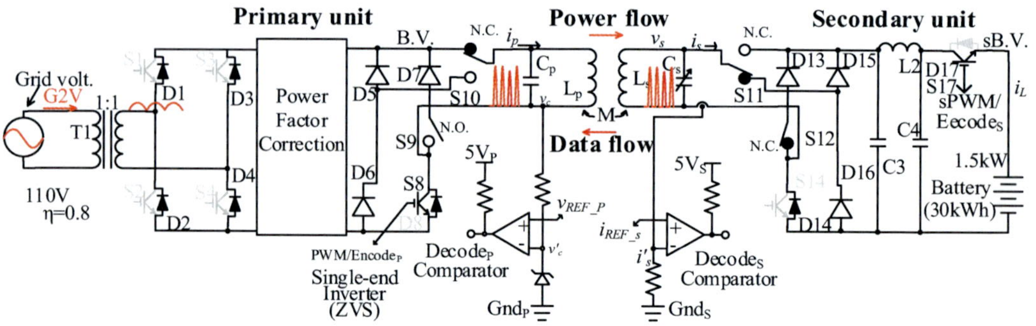

Fig. (8.15). Transceiver configuration in the data attachment method.

In the general data channel, the secondary unit sends a feedback data packet to the primary unit. For example, in Fig. (**8.16**), the feedback data comprises a start bit, b0, and b1. As depicted in Fig. (**8.16**), S17, which is located in the secondary unit, simultaneously controls the output current and modulates the bit "1" during *T*, as expressed in equation 8-9. At points A and B in Fig. (**8.16**), S17 outputs sorting pulses to modulate the start bit and b1, respectively. The longer time (T+Δτ) compared with *T* is the bit "0," as expressed in equation 8.10, where Δτ is a default value.

$$\text{S17modulate/encodedigit"1"=T} \quad \textbf{(8.9)}$$

$$\text{S17modulate/encodedigit"0"=T+}\Delta\tau \quad \textbf{(8.10)}$$

The primary unit uses Decode$_p$ to obtain T (t_{Bit}). The T value for the primary unit is expressed in equations 8.11 and 8.12.

$$T = \frac{1}{4}\sum_{i=1}^{4} T_{Si}, \text{during obtaining star bit} \quad \textbf{(8.11)}$$

$$T = \frac{1}{4}\sum_{i=1}^{4} T_{bit}, \text{ during obtaining data bits} \quad \textbf{(8.12)}$$

The primary unit decodes the bits "1" and "0" as per equations 8.13 and 8.14, respectively. After finding a start bit, the primary unit waits 4.16 ms and then searches the next bit every 8.33 ms. If t_{Bit} satisfies equation 8.14 in each 8.33 ms, this bit is "0"; otherwise, this bit is "1".

$$\text{Secondary unit decodes digit "1"} = t_{Bit} \leq \left(\mathrm{T} + \frac{\Delta\tau}{2}\right) \tag{8.13}$$

$$\text{Secondary unit decodes digit 0} = t_{Bit} > \left(\mathrm{T} + \frac{\Delta\tau}{2}\right) \tag{8.14}$$

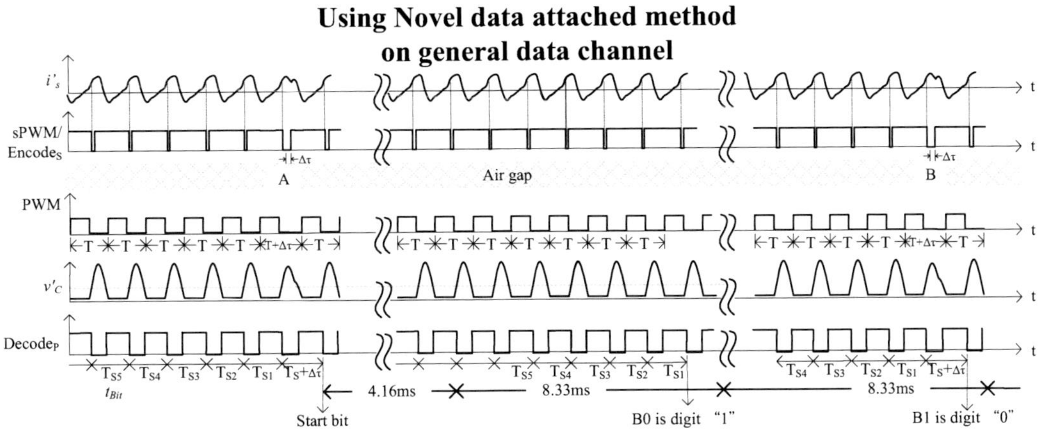

Fig. (8.16). Timing diagrams for the feedback data in the general channel in the data attachment method.

An integrated emergency feedback data packet is displayed in Fig. (**8.17**).

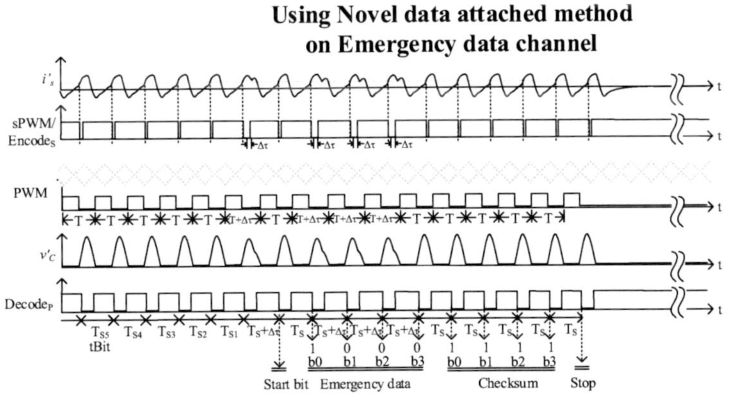

Fig. (8.17). Timing diagrams for the feedback of overcurrent status through the emergency channel in the data attachment method.

CONCLUSION

This chapter proposes an integrated method for coherent wireless power and data transfer. The method features tight construction, switches to control the power flow, handshaking communication before power transfer, current trimming and data attachment after power transfer, and the use of two communication channels after power transfer, namely a general channel and an emergency channel. The charging voltage, charging current, and charging power are transferred through the general channel. When the secondary unit of the proposed topology experiences an emergency, such as overvoltage or overcurrent, it sends a signal through the emergency channel to the primary unit instructing it to quickly halt power transfer.

CHAPTER 9

System Integration and Experimentation

Abstract: This chapter describes experiments on multifunctional contactless grid-t--vehicle (G2V) power flow and bidirectional outer and inner communication based on the giant magnetoresistance (GMR) effect for car parking guidance. In wireless power transfer, an inverter is used to control the G2V power flow in accordance with the zero-voltage switching method for achieving high power transmission efficiency and data transfer without using any radiofrequency devices. This inverter functions over an output power of 400–700 W and an operation frequency of 19–25 kHz. In an experiment on the power transmission quality, the total harmonic distortion in the fundamental frequency was only 23.8% of the significant harmonic frequency, including the DC, second-order, and third-order frequencies, when the output power was 700 W. The data transfer performance when using the current trimming and data attachment methods was examined to generate information quality reports. These reports included information on the lateral offset, output power, and air gap. The information quality was greater than 80% when the lateral offset was −20 to 20 mm and $\Delta\tau$ was 3–5 μs. This lateral offset range (40 mm) was twice the width of the magnetic tape array 3 for GMR guidance, which is sufficient for wireless power charging. When the output power was 400–700 W, and $\Delta\tau$ was 3–5 μs, the information quality was greater than 85%. When the air gap between two power pads was 15–30 mm, and $\Delta\tau$ was 3–5 μs, the information quality was greater than 80%.

Keywords: Current trimming method, Data attachment method, Giant magnetoresistance (GMR), Radio-frequency devices, Zero-voltage switching.

9.1. INTRODUCTION

Experiments were conducted on a prototype electric vehicle (EV) with a giant magnetoresistance (GMR) guidance system [235] to simulate power transfer in an EV parked in a parking lot (Fig. **9.11**). Bidirectional data transfer [231, 232, 255] occurs between the primary and secondary units under grid-to vehicle (G2V) power flow. The GMR guidance system of the prototype EV includes a cell phone that aligns two power pads for ensuring higher power transmission efficiency and information quality when the EV is in a parking lot.

Chih-Cheng Huang & Chun-Liang Lin

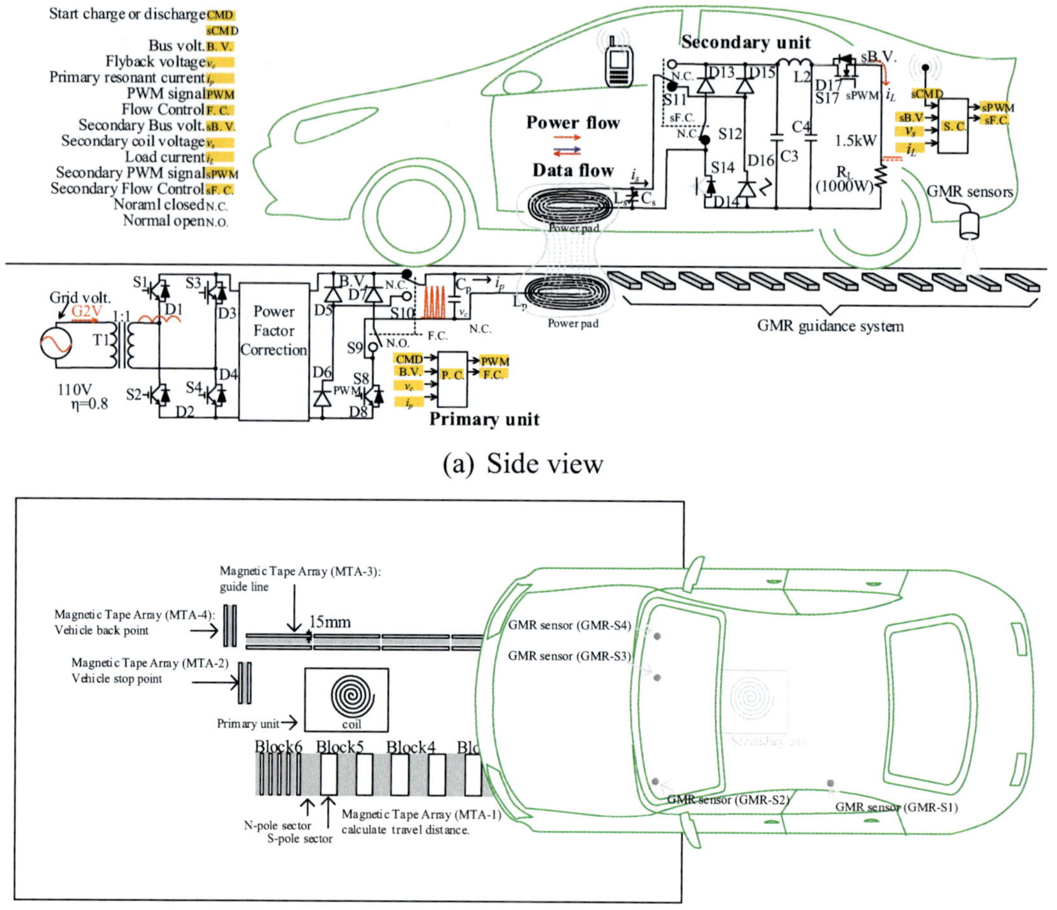

(a) Side view

(b) Top view

Fig. (9.1). Coherent wireless power charging and data transfer for an EV with a GMR guidance system that is parked in a parking lot.

The experimental environment (Fig. **9.2**) was located in the Bioinformatic Computing & Control Lab of the Department of Electrical Engineering, National Chung Hsing University, Taichung, Taiwan [257]. This laboratory supported all the development tools, components, and measurement equipment employed in the experiments.

(a) Right side

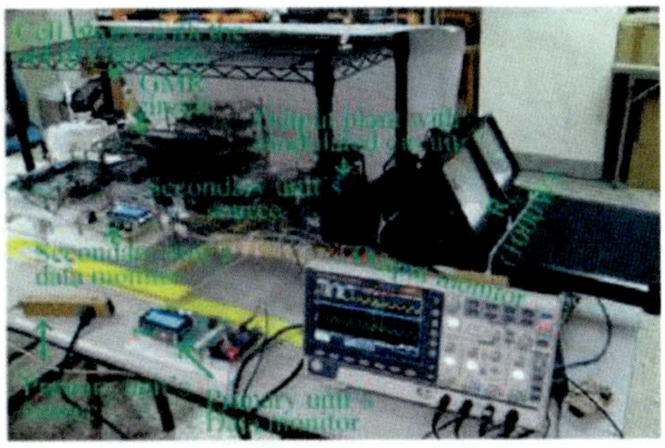

(b) Left side

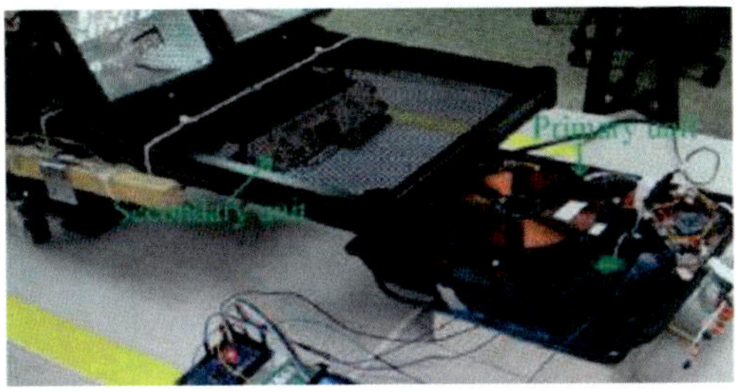

(c) Front side

(d) Rear side

Fig. (9.2). Experimental environment and prototype EV.

The prototype EV could perform coherent wireless power and data transfer, as shown in Fig. (**9.3**). When the prototype EV was parked, its two power pads were aligned within an acceptable lateral offset by using a GMR mobile application. The prototype obtained insufficient battery capacity through handshaking communication. Therefore, G2V power flow occurred during battery charging. In this research, the battery was replaced by two lamps for convenience. The EV's identification number, charging voltage, charging current, and output power from the general data channel were presented on two liquid crystal displays. The EV's identification number was NB-4385; the charging voltage was 141.74 V; the charging current was 3.05 A; the output power was 432.98 W.

(a) (b)

Fig. (9.3). Coherent wireless power and data transfer by the prototype EV.

9.2. Configuration of the Prototype EV

Photographs of the primary and secondary units of the prototype EV are displayed in Fig. (**9.4**), and the parameters of these units are listed in Table **9.1**. The aforementioned units contained an ARM Cortex-M0 CPU for human–computer interaction, power transmission, and information encoding and decoding. The operational frequency of the CPU was 24 MHz, which was sufficient for meeting the present requirements. A communication flowchart of the prototype EV is presented in Fig. (**9.5**).

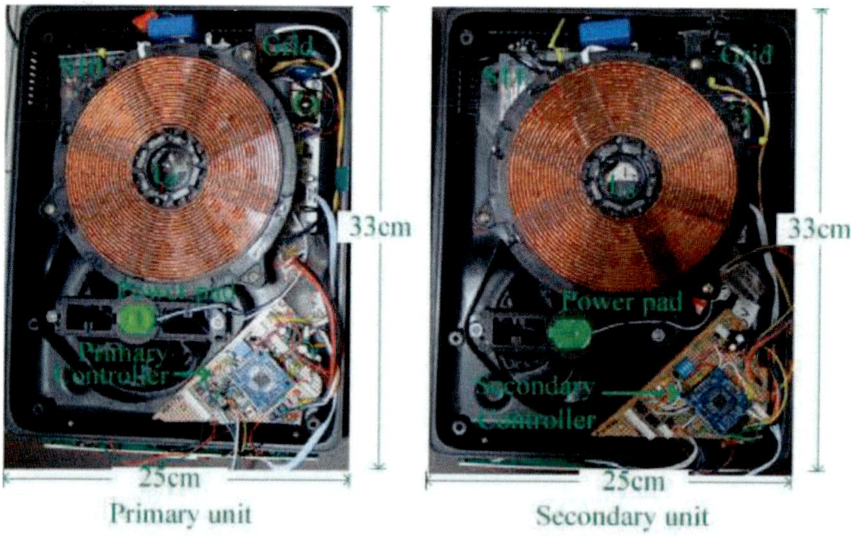

Fig. (9.4). Primary and secondary units of the prototype EV.

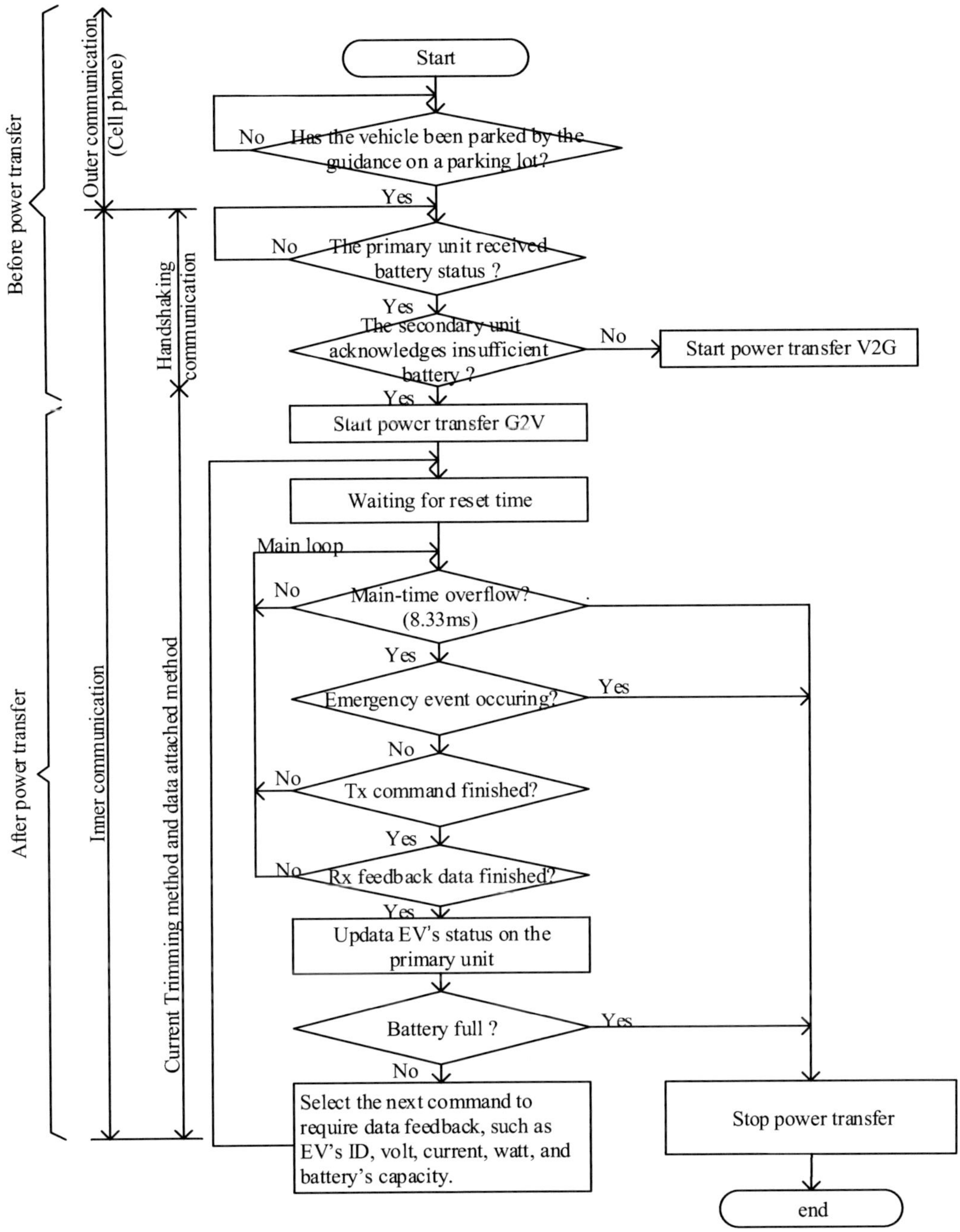

Fig. (9.5). Communication flowchart of the prototype EV.

Table 9.1. Parameters And Specifications Of The Prototype System.

Nominal frequency	22 kHz
Max output power	700 W
Rated load	40 Ω
Primary inductance	69 μH
Primary capacitance	0.35 μF
Primary outer diameter	165 mm
Primary windings	20 turns
Mutual inductance	26.1 μH
Secondary inductance	69.5 μH
Secondary capacitance	0.35 μF
Secondary outer diameter	165 mm
Secondary windings	20 turns
L2	4 μH
C3	280 μF
C4	280 μF
GMR Magnetic Tape width	15mm

9.3. Experimental Verification

Coherent wireless power and data transfer occur in two stages in the constructed EV. First, before power transfer, the power transfer system checks the battery capacity to decide whether G2V or V2G power flow should be conducted through handshaking communication. Second, after power transfer, this system monitors the safety and efficiency of the power flow by using the current trimming and data attachment methods in the general and emergency communication channels. Experiments were conducted in this research to analyze the power transfer quality and communication transfer quality for G2V power transmission.

9.3.1 Experiment on G2V power flow

After the power transfer stage, the inverter in the primary unit generated a high-frequency current and magnetic field from the power supplied by a 110-V AC source with a frequency of 60 Hz. Fig. (**9.6**) displays the measured waveforms for an output power of 400–700 W flowing across the load R_L without data transfer. The aforementioned figures display four types of curves, namely those for i_s, the current in LC tank of secondary side, the gate voltage of S17(Q2) (*i.e.*, the sPWM signal (secondary Pulse Width Modulation), the duty cycle of which is 100%), v_c, the resonant voltage of primary side, and the gate voltage of S8(Q1) [*i.e.*, the pulse width modulation (PWM) signal] under an air gap of 25 mm, a lateral offset of 0 mm and power values of 400, 550, and 700 W.

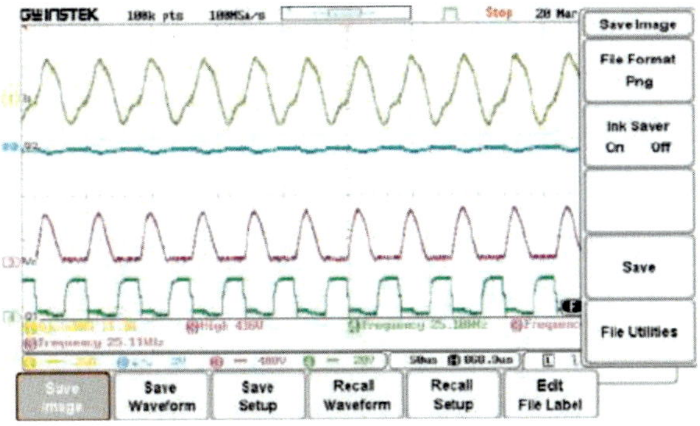

(a) Measured waveforms for a transferred power of 400 W

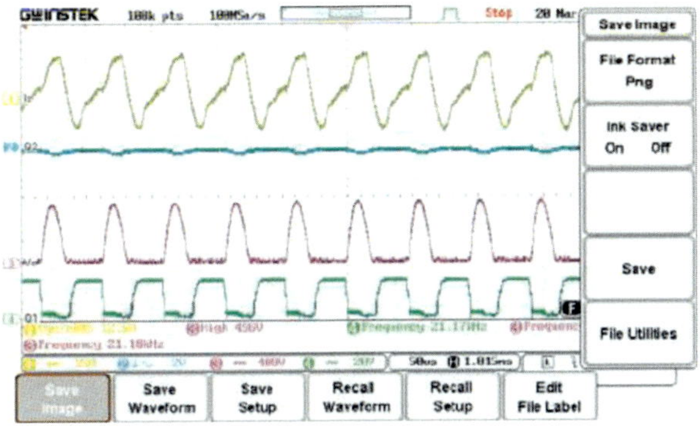

(b) Measured waveforms for a transferred power of 550 W

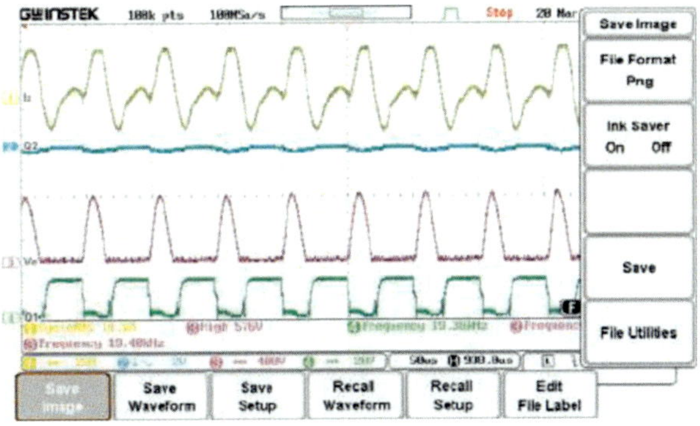

(c) Measured waveforms for a transferred power of 700 W.

Fig. (9.6). Measured waveforms during the wireless transfer of **(a)** 400, **(b)** 550, and **(c)** 700 W of power under an air gap of 25 mm and a lateral offset of 0 mm.

9.3.2. Current Trimming Experiment

The power transfer system of the EV conducts inner communication through two channels, a general channel and an emergency channel. The general channel is used to transfer information on the EV's status, such as the EV's identification number, charging voltage, charging current, and output power. The aforementioned data are fed to the primary unit through a suitable command by using the current trimming method.

In the current trimming method, the PWM signal is shortened by $\Delta\tau$ to modulate the bit "0." The period of the PWM signal for the bit "0" is shorter than that for the bit "1." Fig. (**9.7**) depicts one bit "0" in the general communication channel. Fig. (**9.7**) displays four types of waveforms, namely those for i_s, the Decode$_s$ data of the secondary unit, v_c, and the gate voltage of S8 (*i.e.*, PWM signal). In the center of (Fig. **9.7**), the primary unit modulates the bit '0'with a $\Delta\tau$ value of 4 μs, and the secondary unit receives this bit by using the Decode$_s$ port.

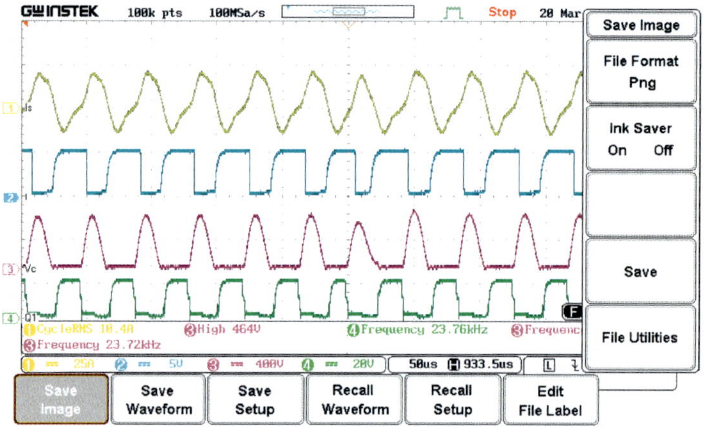

(a) Waveforms for the general channel's Tx command in the coherent wireless transfer of 400 W of power

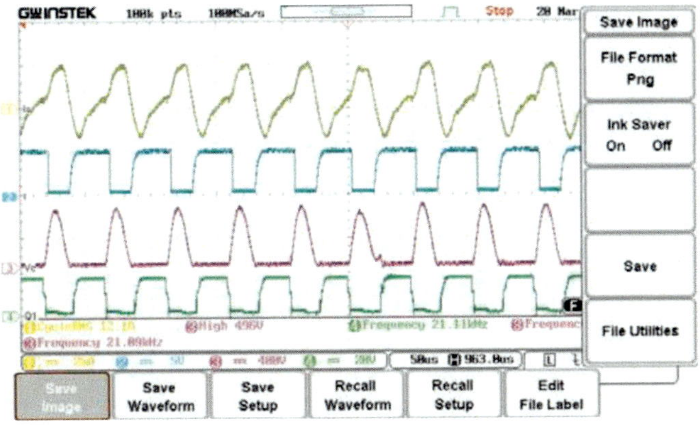

(b) Waveforms for the general channel's Tx command in the coherent wireless transfer of 550 W of power

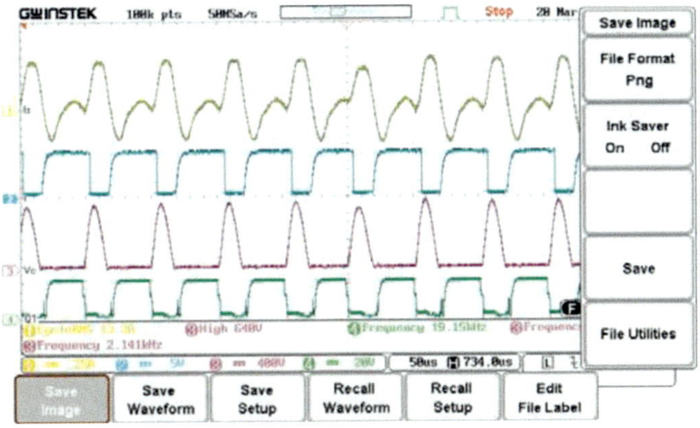

(c) Waveforms for the general channel's Tx command in the coherent wireless transfer of 700 W of power

Fig. (9.7). Measured waveforms for the general channel's Tx command in the coherent wireless transfer of **(a)** 400, **(b)** 550, and **(c)** 700 W of power under an air gap of 25 mm and a lateral offset of 0 mm.

9.3.3. Data Attachment Experiment

The secondary unit's feedback data are modulated during inner communication by using the data attachment method in the general or emergency channel. In the general channel, the sPWM signal is modulated by $\Delta\tau$ to generate the bit "0." The period of the sPWM signal for the bit "0" is longer than that for the bit "1." (Fig. **9.8**) depicts measured waveforms for feedback data transfer in the general channel. This figure depicts four types of waveforms, namely those for i_s, the sPWM signal of the secondary unit, v_c, and the gate voltage of S8 (*i.e.*, the PWM signal).

The secondary unit modulates the bit "0" with a $\Delta\tau$ value of 4 μs, and the primary unit receives this bit from the Decode$_p$ signal. The v_c signal has a longer period for the bit "0" than for the bit "1," which enables this bit to be distinguished. The bit "0" is reflected to the Decode$_p$ port of the primary side.

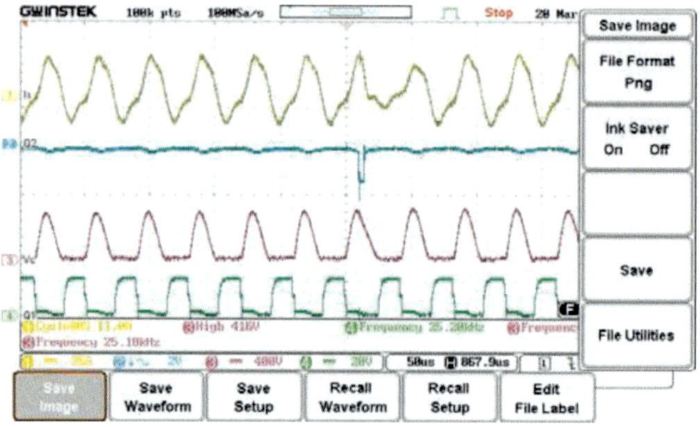

(a) Waveforms for the transfer of the general channel's feedback data in the coherent wireless transfer of 400 W of power

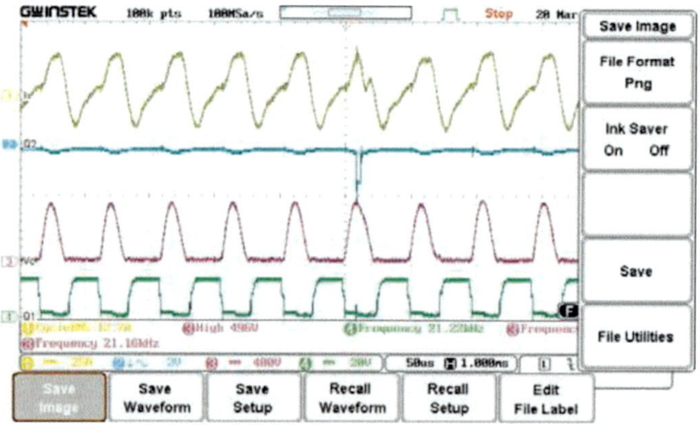

(b) Waveforms for the transfer of the general channel's feedback data in the coherent wireless transfer of 550 W of power

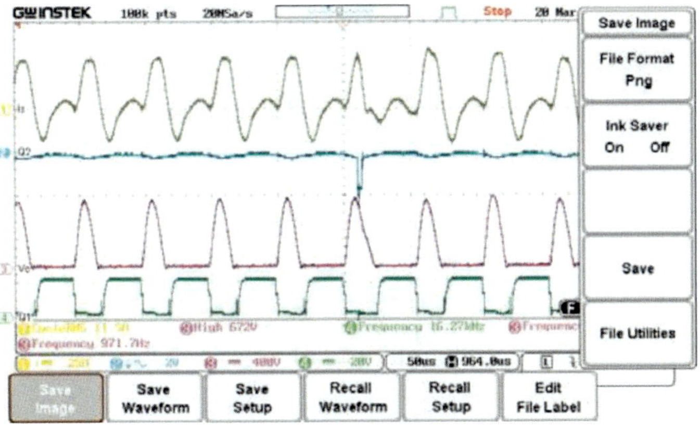

(c) Waveforms for the transfer of the general channel's feedback data in the coherent wireless transfer of 700 W of power

Fig. (9.8). Measured waveforms for the transfer of the general channel's feedback data during the coherent wireless transfer of **(a)** 400, **(b)** 550, and **(c)** 750 W under an air gap of 25 mm and a lateral offset of 0 mm.

9.3.4. Experiment for an Emergency Event

Fig. (**9.9**) displays a data packet that contains information on the charging current of the secondary unit. This packet was sent from the secondary unit to the primary unit through the general channel. The data "0b00110110" in Fig. (**9.9a**) indicates that the secondary unit's charging current was 3.06 A. The meter depicted at the lower left side of this figure confirmed the charging current (i_s). The meter depicted at the lower right side of the aforementioned figure displayed the bus voltage (138.0 V). The transmission time for the aforementioned packet in the general channel was 150 ms. When the secondary unit detected an overcurrent, it transmitted information regarding this event to the primary unit over the emergency channel (Fig. **9.9b**). As displayed in Fig. (**9.9b**), the emergency event occurred at the general data's b3 of the checksum, and the relevant meters displayed a current of 0 A and voltage of 0 V because the primary unit halted power transfer immediately upon receiving the emergency message. The emergency data packet for the overcurrent event is displayed in Fig. (**9.9c**). This packet required a transmission time of 460 μs.

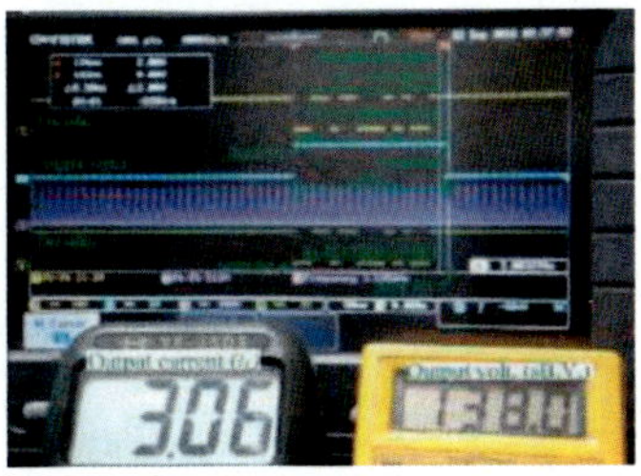

(a)

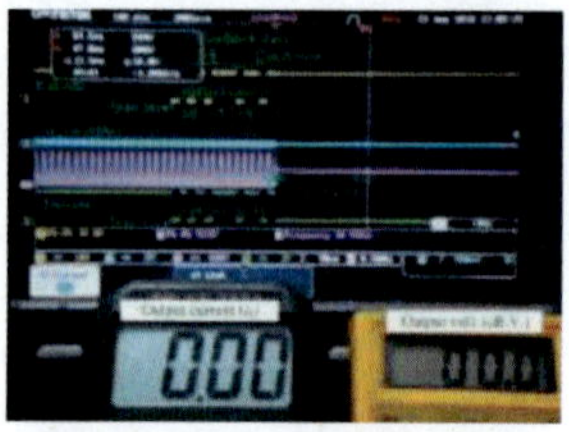

(b)

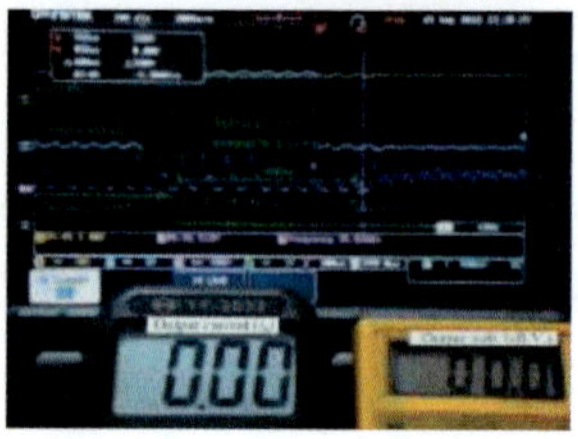

(c)

Fig. (9.9). (a) Data packet containing information on the charging current of the secondary unit; **(b)** stoppage of power transfer within 8.33 ms when the primary unit received information regarding the occurrence of an overcurrent event; **(c)** emergency data packet for the overcurrent event, which was transferred over the emergency channel.

9.3.5. Variations in Δτ and the Power Transfer Efficiency Under Data Attachment

Fig. (**9.10**) shows the variation in $\Delta\tau$ with the duty cycle of the PWM signal. After adjusting the duty cycle of S17 from 94% to 85%, the following near-linear relationship was obtained in the Decode_p port.

$$\text{Primary unit decodes digit } 0 = t_{Bit} > (\text{T} + \Delta\tau + \text{offset time}) \quad \textbf{(1)}$$

Where t_{Bit} is the feedback period. The offset time was −1 μs in the experiment.

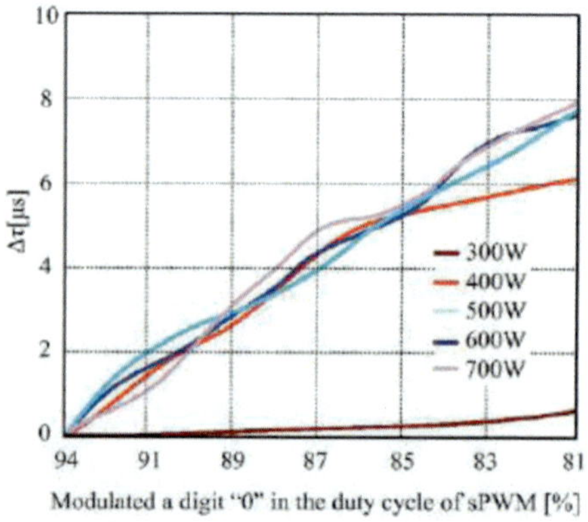

Fig. (9.10). Variation in $\Delta\tau$ with the duty cycle of the sPWM signal (S17).

Fig. (**9.11**) illustrates the overall power transfer efficiency with general communication. The power transfer efficiency decreased marginally when data feedback 2 bytes to the primary side in the general data channel every 0.5 s.

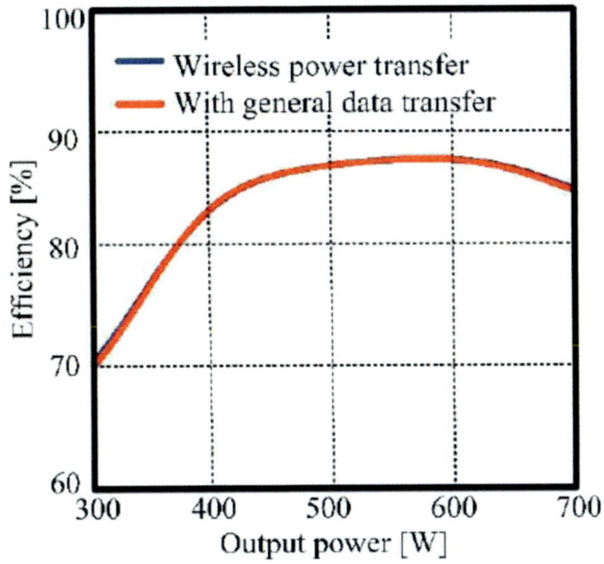

Fig. (9.11). A marginal decrease in the transmission efficiency during coherent wireless power and data transfer.

9.3.6. Power Transmission Quality

The power transmission system of the prototype EV contained an inductor–capacitor (LC) tank with the parallel-parallel (*PP)* type for the linking of magnetic fields and the transfer of power and data through an air gap. The LC tank filtered most harmonic elements when the inverter operated in the zero-voltage switching mode and switched the current frequently. The harmonic spectrum of the secondary current (i_s) under an output power of 700 W *is displayed* in Fig. (**9.12**). The total harmonic distortion for the fundamental frequency was only 23.8% of the significant harmonic frequency, including the DC, second-order, and third-order frequencies when the output power was 700 W. Moreover, the power transmission system employed a load-detecting function in the handshaking communication stage to detect loads for ensuring that energy was fully transmitted to the secondary unit. Therefore, high-order harmonics were considerably minimized in the aforementioned system.

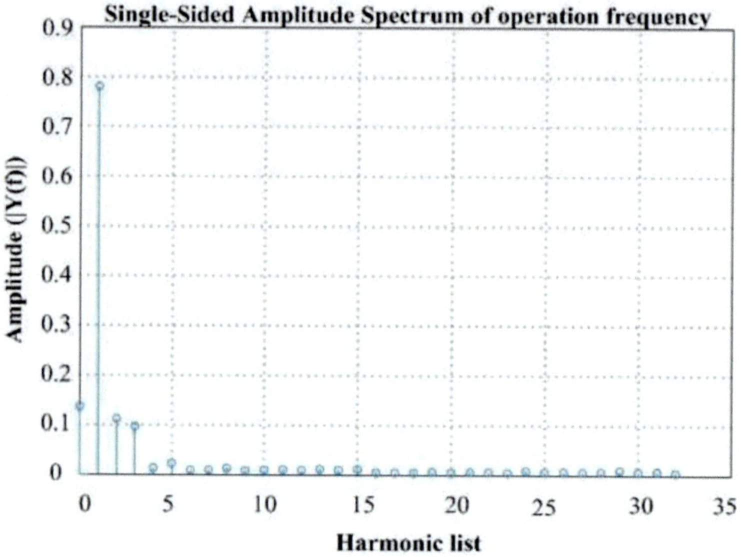

Fig. (9.12). Harmonic of i_s for the wireless power transfer output power 700W.

To reduce current variation during battery charge, a second-order low-pass filter was added to the secondary unit. Fig. (**9.13**) depicts the waveforms obtained for i_s, the sPWM signal (S17) of the secondary unit, the bus voltage of the secondary unit (approximately 185 V), and i_L (approximately 3.1 A) under an output power of 700 W after filtering using the aforementioned filter. The current i_L and bus voltage are smooth when S17 modulated data.

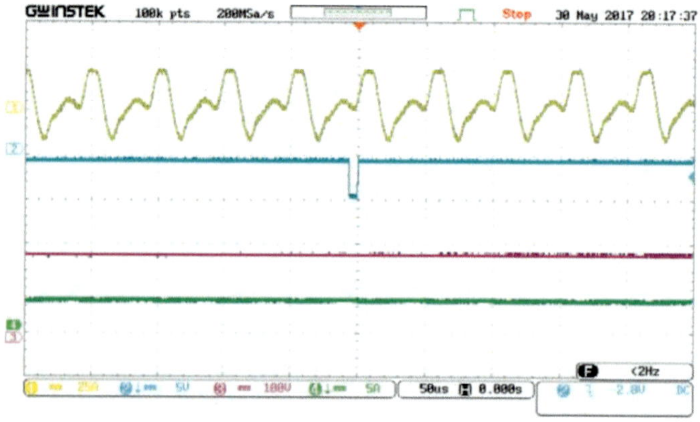

Fig. (9.13). Measured waveforms obtained for the sPWM signal (S17) of the secondary unit, the bus voltage of the secondary unit, and i_L under coherent wireless power and data transfer with an output voltage of 700 W when modulating a data bit with the power output filtered using a second-order low-pass filter.

9.3.7. Data Transmission Quality

Experiments were conducted to verify the information quality under a fixed lateral offset, varying output power, and varying air gap. The information quality was calculated as follows:

$$\text{communication quality} = \frac{pass\ count}{total\ count}\% \tag{2}$$

where the total count was 1000 for the general data transmitted in the general channel and the pass count was the number of correct data packets received, which is expressed as follows:

$$\text{Checksum} = \text{Data } 2's \tag{3}$$

If the checksum was held, the data received was recognized as correct.

The experimental results obtained for a lateral shift of L_p and L_s are depicted in Fig. (**9.14a**). The information quality was higher than 80% when the lateral offset was −20 to 20 mm and $\Delta\tau$ was 3–5 μs. Fig. (**9.14b**) illustrates the effect of the output power on information quality. The information quality was greater than 85% when the output power was varied from 400 to 700 W and $\Delta\tau$ was 3–5 μs.

The information quality was greater than 80% when the air gap between the two power pads varied from 15 to 30 mm, and $\Delta\tau$ was 3–5 μs.

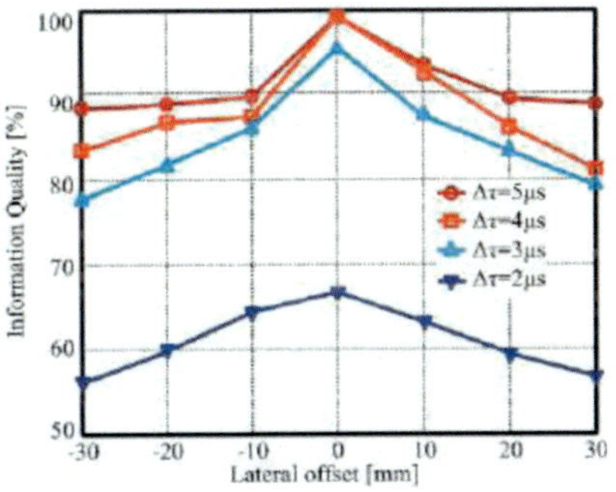

(a)

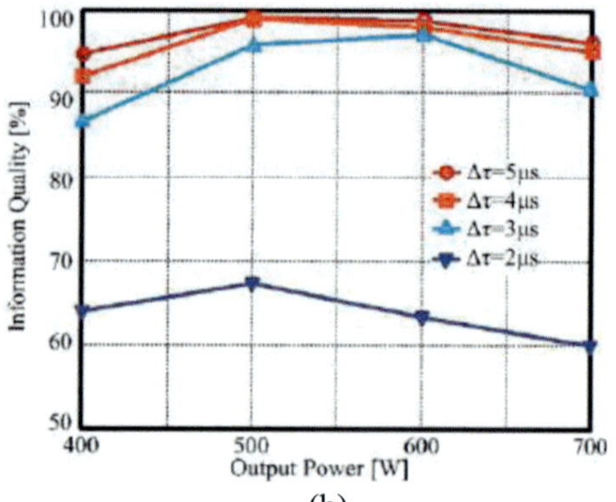

(b)

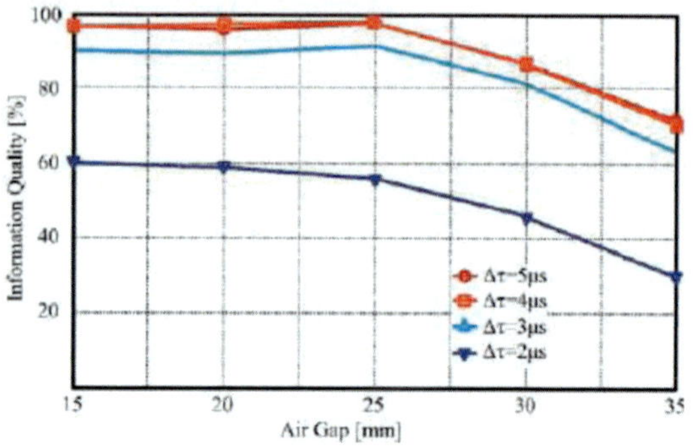

(c)

Fig. (9.14). **(a)** Measured information quality under **(a)** different lateral offsets between L_p and L_s with an air gap of 25 mm, **(b)** different output powers, and **(c)** different air gaps between L_p and L_s

CONCLUSION

The functioning of a prototype EV with a GMR guidance system in a parking space was simulated in this research. A cell phone was used to align the power transmission pad under the vehicle's chassis with the grounded power pad. The prototype conducted handshaking communication to link itself to the grid before conducting power transfer. The default power flow mode was the G2V mode. Experiments were conducted on the G2V power transfer of the prototype EV within an output power range of 300–700W, the current trimming function, the data attachment method, emergency event verification, the transmission efficiency, the total harmonic distortion for the fundamental frequency at the output power of 700 W, and the data transmission quality. The experimental results proved the feasibility of the designed prototype.

REFERENCES

[1] J.Y. Yong, V.K. Ramachandaramurthy, and K.M. Tan, "A review on the state-of-the-art technologies of electric vehicle, its impacts and prospects", *Renew. Sustain. Energy Rev.,* vol. 49, pp. 365-385 .[http://dx.doi.org/10.1016/j.rser.2015.04.130]

[2] A.Y. Saber, and G.K. Venayagamoorthy, "One million plug-in electric vehicles on the road by 2015", IEEE Intell, Ed., *Syst. Conf,* pp. 141-147 .[http://dx.doi.org/10.1109/ITSC.2009.5309691]

[3] *Adoption of the Paris Agreement,* United Nations: New York, NY, USA., .

[4] World Energy Outlook, Int. Energy Agency, Vienna, Austria, 2011.

[5] P. Y. KONG1, and G. K. KARAGIANNIDIS, "Charging Schemes for Plug-In Hybrid Electric Vehicles in Smart Grid: A Survey", IEEE Access, vol. 4, pp. 6846-6875, 2016.[http://dx.doi.org/10.1109/ACCESS.2016.2614689]

[6] F. Kong, and X. Liu, "Sustainable Transportation with Electric Vehicles", *Foundations and Trends in Electric Energy Systems,* vol. 2, no. 1, pp. 1-132.[http://dx.doi.org/ 10.1561/3100000016]

[7] A. Emadi, M. Ehsani, and J.M. Miller, *Vehicular Electric Power Systems: Land, Sea, Air, and Space Vehicles.,* Marcel Dekker: New York, .[http://dx.doi.org/10.1201/9780203913468]

[8] C.C. Chan, "The state of the art of electric hybrid and fuel cell vehicles", *Proc. IEEE,* vol. 95, no. 4, pp. 704-718 .[http://dx.doi.org/ 10.1109/JPROC.2007.892489]

[9] https://ihsmarkit.com/index.html

[10] https://www.toyota.com/prius/

[11] M. Ehsani, Y. Gao, and J.M. Miller, "Hybrid Electric Vehicles: Architecture and motor drives", *Proc. IEEE,* vol. 95, no. 4, pp. 719-728.[http://dx.doi.org/10.1109/JPROC.2007.892492]

[12] A. Emadi, K. Rajashekara, S.S. Willaimson, and S.M. Lukic, "Topological overview of hybrid electric and fuel cell vehicular power systems architectures and configurations", *IEEE Trans. Vehicular Technol.,* vol. 54, no. 3, pp. 763-770.[http://dx.doi.org/10.1109/TVT.2005.847445]

[13] M. Ehsani, Y. Gao, S. Gay, and A. Amade, "Modern Electric, Hybrid Electric, and Fuel Cell Vehicles: Fundamentals, Theory, and Design", *Boca Raton, FL: CRC,* .

[14] J. M. Miller, "Hybrid electric vehicle propulsion system architecture of the e-CVT type", *IEEE Trans. Power Electron,* vol. 21, no. 3, pp. 756-767.[http://dx.doi.org/10.1109/TPEL.2006.872372]

[15] J.M. Miller, M. Ehsani, and Y. Gao, "Understanding power flows in HEV eCVT's with ultracapacitor boosting using simplorer", *IEEE Power Propulsion Conf., Illinois Inst. Technol.*Chicago, IL

[16] Y. Cheng, K. Chen, C.C. Chan, A. Bouscayrol, S. Cui, and A. Berthon, "Global modeling and control strategy simulation for a Hybrid Electric Vehicle using electrical variable transmission", *IEEE Veh. Technol. Mag.,* vol. 4, no. 2, pp. 73-79.[http://dx.doi.org/10.1109/MVT.2009.932541]

[17] M.J. Hoeijmakers, and J.A. Ferreira, "The electric variable transmission", *IEEE Trans. Ind. Appl.,* vol. 42, no. 4, pp. 1092-1100.[http://dx.doi.org/10.1109/TIA.2006.877736]

[18] Y. Cheng, S. Cui, L. Song, and C.C. Chan, "The study of the operation modes and control strategies of an advanced electromechanical converter for automobiles", *IEEE Trans. Magn.,* vol. 43, no. 1, pp. 430-433.[http://dx.doi.org/10.1109/TMAG.2006.887649]

[19] K.T. Chau, C.C. Chan, and C. Liu, "Overview of permanent-magnet brushless drives for electric and Hybrid Electric Vehicles", *IEEE Trans. Ind. Electron.,* vol. 55, no. 6, pp. 2246-2257.[http://dx.doi.org/10.1109/TIE.2008.918403]

[20] K. Rajashekara, "Present status and future trends in electric vehicle propulsion technologies", *IEEE J.*

Chih-Cheng Huang & Chun-Liang Lin

Emerg. Sel. Top. Power Electron., vol. 1, no. 1, pp. 3-10.[http://dx.doi.org/10.1109/JESTPE.2013.2259614]

[21] E.A. Grunditz, and T. Thiringer, "Performance Analysis of Current BEVs Based on a Comprehensive Review of Specifications", *IEEE Trans. Electrification,* vol. 2, no. 3, pp. 270-289.[http://dx.doi.org/10.1109/TTE.2016.2571783]

[22] X. Hu, C. Zou, X. Tang, T. Liu, and L. Hu, "Cost-optimal energy management of hybrid electric vehicles using fuel cell/battery health-aware predictive control", *IEEE Trans. Power Electron.,* vol. 35, no. 1, pp. 382-392.[http://dx.doi.org/10.1109/TPEL.2019.2915675]

[23] H. Zheng, J. Wu, W. Wu, and Y. Wang, "Integrated Motion and Powertrain Predictive Control of Intelligent Fuel Cell/Battery Hybrid Vehicles", *IEEE Trans. Industr. Inform.,* vol. 16, no. 5, pp. 3397-3406.[http://dx.doi.org/10.1109/TII.2019.2956209]

[24] J. Bauman, and M. Kazerani, "A comparative study of fuel-cell–battery fuel-cell–ultracapacitor and fuel-cell–battery–ultracapacitor vehicles", *IEEE Trans. Vehicular Technol.,* vol. 57, no. 2, pp. 760-769.[http://dx.doi.org/10.1109/TVT.2007.906379]

[25] C. Bao, M. Ouyang, and B. Yi, "Analysis of water management in proton exchange membrane membrane fuel cells", *Tsinghua Sci. Technol.,* vol. 11, no. 1, p. •••.[http://dx.doi.org/10.1016/S1007-0214(06)70155-9]

[26] A. Emadi, Y.J. Lee, and K. Rajashekara, "Power electronics and motor drives in electric hybrid electric and plug-in hybrid electric vehicles", *IEEE Trans. Ind. Electron.,* vol. 55, no. 6, pp. 2237-2245.[http://dx.doi.org/10.1109/TIE.2008.922768]

[27] Z. Liu, Q. Wu, S. Huang, L. Wang, and Y. Xue, "Optimal Day-Ahead Charging Scheduling of Electric Vehicles Through an Aggregative Game Model", *IEEE Trans. Smart Grid,* vol. 9, no. 5, pp. 5173-5184.[http://dx.doi.org/10.1109/TSG.2017.2682340]

[28] T. Zhao, and Z. Ding, "Distributed Initialization-Free Cost-Optimal Charging Control of Plug-In Electric Vehicles for Demand Management", *IEEE Trans. Industr. Inform.,* vol. 13, no. 6, pp. 2791-2801.[http://dx.doi.org/10.1109/TII.2017.2685422]

[29] K. M. Muttaqi, M. R. Islam, and D. Sutanto, "Future Power Distribution Grids: Integration of Renewable Energy, Energy Storage, Electric Vehicles, Superconductor, and Magnetic Bus", *IEEE Trans. App. Superconductivity,* vol. 29, no. 2, .[http://dx.doi.org/10.1109/TASC.2019.2895528]

[30] H. Turker, and S. Bacha, "Optimal minimization of plug-in electric vehicle charging cost with vehicle-to-home and vehicle-to-grid concepts", *IEEE Trans. Vehicular Technol.,* vol. 67, no. 11, pp. 10281-10292.[http://dx.doi.org/10.1109/TVT.2018.2867428]

[31] V. Monteiro, B. Exposto, J.C. Ferreira, and J.L. Afonso, "Improved vehicle-to-home (V2H) operation mode: Experimental analysis of the electric vehicle as off-line UPS", *IEEE Trans. Smart Grid,* vol. 8, no. 6, pp. 2702-2711.[http://dx.doi.org/10.1109/TSG.2016.2535337]

[32] C. Liu, K.T. Chau, D. Wu, and S. Gao, "Opportunities and challenges of vehicle-to-home vehicle-t--vehicle and vehicle-to-grid technologies", *Proc. IEEE,* vol. 101, no. 11, pp. 2409-2427.[http://dx.doi.org/10.1109/JPROC.2013.2271951]

[33] C.C. Chan, A. Bouscayrol, and K. Chen, "Electric hybrid and fuel-cell vehicles: architectures and modeling", *IEEE Trans. Vehicular Technol.,* vol. 59, no. 2, pp. 589-598.[http://dx.doi.org/10.1109/TVT.2009.2033605]

[34] M. Zeraoulia, M. Benbouzid, and D. Diallo, "Electric motor drive selection issues for HEV propulsion systems: A comparative study", *IEEE Trans. Vehicular Technol.,* vol. 55, no. 6, pp. 1756-1764.[http://dx.doi.org/10.1109/TVT.2006.878719]

[35] A.M. Dizqah, B. Lenzo, A. Sorniotti, P. Gruber, S. Fallah, and J. De Smet, "A Fast and Parametric Torque Distribution Strategy for Four-Wheel-Drive Energy-Efficient Electric Vehicles", *IEEE Trans. Ind. Electron.,* vol. 63, no. 7, pp. 4367-4376.[http://dx.doi.org/10.1109/TIE.2016.2540584]

[36] W. Wang, X. Chen, and J. Wang, "Motor/generator applications in electrified vehicle chassis—A survey", *IEEE Trans. Transp. Electrific,* vol. 5, no. 3, pp. 584-601.[http://dx.doi.org/10.1109/TTE.2019.2934340]

[37] P. Ramesh, and N.C. Lenin, "High Power Density Electrical Machines for Electric Vehicles—Comprehensive Review Based on Material Technology", *IEEE Trans. Magn.,* vol. 55, no. 11, .[http://dx.doi.org/10.1109/TMAG.2019.2929145]

[38] "Tesla motors club BEV Spec Comparison with Model 3", [Online].Available: https:// teslamotorsclub.com/tmc/threads/bev-spec-comparison-with-model-3.67336/

[39] J.Y. Wong, *Theory of Ground Vehicle.,* Wiley: New York, pp. 257-258.

[40] M. Ehsani, Y. Gao, and S. Gay, "Characterization of electric motor drives for traction applications", [http://dx.doi.org/10.1109/IECON.2003.1280101]

[41] Z.Q. Zhu, and D. Howe, "Electrical machines and drives for electric, hybrid, and fuel cell vehicles", *Proc. IEEE,* vol. 95, no. 4, pp. 746-765.[http://dx.doi.org/10.1109/JPROC.2006.892482]

[42] S.J. Rind, Y. Ren, Y. Hu, J. Wang, and L. Jiang, "Configurations and Control of Traction Motors for Electric Vehicles: A Review", *Chinese Journal of Electrical Engineering,* vol. 3, no. 3, pp. 1-17.[http://dx.doi.org/10.23919/CJEE.2017.8250419]

[43] M. Ehsani, K. Rahman, and H. Toliyat, "Propulsion system design of electric and hybrid vehicles", *IEEE Trans. Ind. Electron.,* vol. 44, no. 1, pp. 19-27.[http://dx.doi.org/10.1109/41.557495]

[44] Y. Kai, Y. Shi, Y. Xu, and R.D. Lorenz, "A comparative overview of indirect field oriented control(IFOC) and deadbeat direct torque and flux control(DB-DTFC) for AC motor drives", *Chinese Journal of Electrical Engineering,* vol. 1, no. 1, pp. 9-20.[http://dx.doi.org/10.23919/CJEE.2015.7933134]

[45] L. Zhang, Y. Fan, R. Cui, R.D. Lorenz, and M. Cheng, "Fault-tolerant direct torque control of five-phase FTFSCW-IPM motor based on analogous three-phase SVPWM for electric vehicle applications", *IEEE Trans. Vehicular Technol.,* vol. 67, no. 2, pp. 910-919.[http://dx.doi.org/10.1109/TVT.2017.2760980]

[46] Y.N. Tatte, and M.V. Aware, "Torque ripple and harmonic reduction in a three-level inverter-fed direct-torque-controlled five-phase induction motor", *IEEE Trans. Ind. Electron.,* vol. 64, no. 7, pp. 5265-5275.[http://dx.doi.org/10.1109/TIE.2017.2677346]

[47] R.J. Parker, and R.J. Studders, *Permanent Magnets and Their Application.,* Wiley: New York, .

[48] K.T. Chau, Q. Sun, Y. Fan, and M. Cheng, "Torque ripple minimization of doubly salient permanent-magnet motors", *IEEE Trans. Energ. Convers.,* vol. 20, no. 2, pp. 352-358.[http://dx.doi.org/10.1109/TEC.2004.841507]

[49] G. Pellegrino, A. Vagati, P. Guglielmi, and B. Boazzo, "Performance comparison between surface-mounted and interior PM motor drives for electric vehicle application", *IEEE Trans. Ind. Electron.,* vol. 59, no. 2, pp. 803-811.[http://dx.doi.org/10.1109/TIE.2011.2151825]

[50] R.W. Cao, C. Mi, and M. Cheng, "Quantitative Comparison of Flux- Switching Permanent-Magnet Motors with Interior Permanent Magnet Motor for EV, HEV, and PHEV Applications", *IEEE Trans. Magn.,* vol. 48, no. 8, pp. 2374-2384.[http://dx.doi.org/10.1109/TMAG.2012.2190614]

[51] B. Cao, B.M. Grainger, X. Wang, Y. Zou, G.F. Reed, and Z-H. Mao, "Direct torque model predictive control of a five-phase permanent magnet synchronous motor", *IEEE Trans. Power Electron.,* vol. 36, no. 2, pp. 2346-2360.[http://dx.doi.org/10.1109/TPEL.2020.3011312]

[52] T. Song, Z. Zhang, H. Liu, and W. Hu, "Multi-objective optimisation design and performance comparison of permanent magnet synchronous motor for EVs based on FEA", *IET Electr. Power Appl.,* vol. 13, no. 8, pp. 1157-1166.[http://dx.doi.org/10.1049/iet-epa.2019.0069]

[53] A. Fratta, G. Troglia, A. Vagati, and F. Villata, "Evaluation of torque ripple in high performance

synchronous reluctance machines", *Industry Applications Conference, IEEE IAS Annual Meeting, Toronto, ON, Canada,* vol. vol. 1, pp. 163-170.[http://dx.doi.org/10.1109/IAS.1993.298919]

[54] S. Morimoto, "Performance of PM assisted synchronous reluctance motor for high-efficiency and wide constant-power operation", *IEEE Trans. Ind. Appl.,* vol. 37, no. 5, pp. 1234-1240.[http://dx.doi.org/10.1109/28.952497]

[55] T.M. Jahns, and V. Caliskan, "Uncontrolled generator operation of interior PM synchronous machines following high-speed inverter shutdown", *IEEE Trans. Ind. Appl.,* vol. 35, no. 6, pp. 1347-1357.[http://dx.doi.org/10.1109/28.806049]

[56] Z.M. Zhao, S. Meng, C.C. Chan, and E.W.C. Lo, "A novel induction machine design suitable for inverter-driven variable speed systems", *IEEE Trans. Energ. Convers.,* vol. 15, no. 4, pp. 413-420.[http://dx.doi.org/10.1109/60.900502]

[57] T. Wang, P. Zheng, and S. Cheng, "Design characteristics of the induction motor used for hybrid electric vehicle", *IEEE Trans. Magn.,* vol. 41, no. 1, pp. 505-508.[http://dx.doi.org/10.1109/TMAG.2004.838967]

[58] Z. Dong, "Operating point selected flux-weakening control of induction motor for torque-improved high-speed operation under multiple working conditions", *IEEE Trans. Power Electron.,* vol. 34, no. 12, pp. 12011-12023.[http://dx.doi.org/10.1109/TPEL.2019.2905536]

[59] Y. Hu, X. Song, W. Cao, and B. Ji, "New SR drive with integrated charging capacity for plug-in hybrid electric vehicles (PHEVs)", *IEEE Trans. Ind. Electron.,* vol. 61, no. 10, pp. 5722-5731.[http://dx.doi.org/10.1109/TIE.2014.2304699]

[60] I. Boldea, L.N. Tutelea, L. Parsa, and D. Dorrell, "Automotive electric propulsion systems with reduced or no permanent magnets: an overview", *IEEE Trans. Ind. Electron.,* vol. 61, no. 10, pp. 5696-5711.[http://dx.doi.org/10.1109/TIE.2014.2301754]

[61] J.J. Gu, and H. Chen, "Implementation of the three-phase switched reluctance machine system for motors and generators", *IEEE/ASME Trans. Mechatron.,* vol. 15, no. 3, pp. 421-432.[http://dx.doi.org/10.1109/TMECH.2009.2027901]

[62] H.H. Moghbelli, and M.H. Rashid, "Performance review of the switched reluctance motor drives", *34th Midwest Symp. Circuits Syst,* vol. vol. 1, pp. 162-165.

[63] Z. Qianfan, C. Shumei, and T. Xinjia, "Hybrid switched reluctance motor applied in electric vehicles", [http://dx.doi.org/10.1109/VPPC.2007.4544152]

[64] J. Haataja, *A comparative performance study of four-pole induction motors and synchronous reluctance motors in variable speed drives,* .

[65] K.M. Rahman, and S.E. Schulz, "Design of high-efficiency and hightorque-density switched reluctance motor for vehicle propulsion", *IEEE Trans. Ind. Appl.,* vol. 38, no. 6, pp. 1500-1507.[http://dx.doi.org/10.1109/TIA.2002.805571]

[66] X.D. Xue, K.W.E. Cheng, T.W. Ng, and N.C. Cheung, "Multiobjective optimization design of in-wheel switched reluctance motors in electric vehicles", *IEEE Trans. Ind. Electron.,* vol. 57, no. 9, pp. 2980-2987.[http://dx.doi.org/10.1109/TIE.2010.2051390]

[67] K. Kiyota, and A. Chiba, "Design of switched reluctance motor competitive to 60-kW IPMSM in third-generation hybrid electric vehicle", *IEEE Trans. Ind. Appl.,* vol. 48, no. 6, pp. 2303-2309.[http://dx.doi.org/10.1109/TIA.2012.2227091]

[68] Y. Hu, C. Gan, W. Cao, and S.J. Finney, "Flexible Fault-Tolerant Topology for Switched Reluctance Motor Drives", *IEEE Trans. Power Electron.,* vol. 31, no. 6, pp. 4654-4668.[http://dx.doi.org/10.1109/TPEL.2015.2477165]

[69] J.d. Santiago, H. Bernhoff, B. Ekergård, S. Eriksson, S. Ferhatovic, R. Waters, and M. Leijon, "Electrical Motor Drivelines in Commercial All-Electric Vehicles: A Review", *IEEE Trans. Vehicular Technol.,* vol. 61, no. 2, pp. 475-484.[http://dx.doi.org/10.1109/TVT.2011.2177873]

[70] E. Martinez-Laserna, "Technical viability of battery second life: a study from the ageing perspective", *IEEE Trans. Industry App,* .[http://dx.doi.org/10.1109/TIA.2018.2801262]

[71] D.N.T. How, M.A. Hannan, M.S.H. Lipu, and P.J. Ker, "State of charge estimation for lithium-ion batteries using model-based and data-driven methods: A review", *IEEE Access,* vol. 7, pp. 136116-136136.[http://dx.doi.org/10.1109/ACCESS.2019.2942213]

[72] M.A. Roscher, O.S. Bohlen, and D.U. Sauer, "Reliable state estimation of multicell lithium-ion battery systems", *IEEE Trans. Electron. Comput.,* vol. 26, pp. 737-743.

[73] H.E. Perez, X. Hu, S. Dey, and S.J. Moura, "Optimal charging of Li-Ion batteries with coupled electro-thermal-aging dynamics", *IEEE Trans. Vehicular Technol.,* vol. 66, no. 9, pp. 7761-7770.[http://dx.doi.org/10.1109/TVT.2017.2676044]

[74] R. Xiong, H. He, F. Sun, and K. Zhao, "Evaluation on state of charge estimation of batteries with adaptive extended Kalman filter by experiment approach", *IEEE Trans. Vehicular Technol.,* vol. 62, no. 1, pp. 108-117.[http://dx.doi.org/10.1109/TVT.2012.2222684]

[75] R. Xiong, F. Sun, H. He, and T. Nguyen, "A data-driven adaptive state of charge and power capability joint estimator of lithium-ion polymer battery used in electric vehicles", *Energy,* vol. 63, pp. 295-308.[http://dx.doi.org/10.1016/j.energy.2013.10.027]

[76] M. Swierczynski, D-I. Stroe, A.I. Stan, R. Teodorescu, and S.K. Kær, "Lifetime Estimation of the Nanophosphate LiFePO4/C Battery Chemistry Used in Fully Electric Vehicles", *IEEE Trans. Ind. Appl.,* vol. 51, no. 4, pp. 3453-3461.[http://dx.doi.org/10.1109/TIA.2015.2405500]

[77] J. Jaguemont, L. Boulon, P. Venet, Y. Dube, and A. Sari, "Lithium-ion battery aging experiments at subzero temperatures and model development for capacity fade estimation", *IEEE Trans. Vehicular Technol.,* vol. 65, no. 6, pp. 4328-4343.[http://dx.doi.org/10.1109/TVT.2015.2473841]

[78] M.T. Sougrati, A. Darwiche, X. Liu, A. Mahmoud, R.P. Hermann, S. Jouen, L. Monconduit, R. Dronskowski, and L. Stievano, "Transition-metal carbodiimides as molecular negative electrode materials for lithium- and sodium-ion batteries with excellent cycling properties", *Angew. Chem. Int. Ed. Engl.,* vol. 55, no. 16, pp. 5090-5095.[http://dx.doi.org/10.1002/anie.201600098] [PMID: 26989882]

[79] C. Spanos, D.E. Turney, and V. Fthenakis, "Life-cycle analysis of flow assisted nickel zinc-, manganese dioxide-, and valve-regulated lead-acid batteries designed for demand-charge reduction", *Renew. Sustain. Energy Rev.,* vol. 43, pp. 478-494.[http://dx.doi.org/10.1016/j.rser.2014.10.072]

[80] M. Garcia-Plaza, D. Serrano-Jimenez, J.E.G. Carrasco, and J. Alonso-Martinez, "A Ni-Cd battery model considering state of charge and hysteresis effects", *J. Power Sources,* vol. 275, pp. 595-604.[http://dx.doi.org/10.1016/j.jpowsour.2014.11.031]

[81] Y. Zhu, W. H. H. Zhu, Z. Davis, and B. J. Tatarchuk, "Simulation of Ni-MH batteries via an equivalent circuit model for energy storage applications", *Adv. Phys. Chem,* .[http://dx.doi.org/10.1155/2016/4584781]

[82] Q. Lai, H. Zhang, X. Li, L. Zhang, and Y. Cheng, "A novel single flow zinc bromine battery with improved energy density", *J. Power Sources,* vol. 235, pp. 1-4.[http://dx.doi.org/10.1016/j.jpowsour.2013.01.193]

[83] A. Fotouhi, D.J. Auger, K. Propp, S. Longo, and M. Wild, "A review on electric vehicle battery modelling: From Lithium-ion toward Lithium-Sulphur", *Renew. Sustain. Energy Rev.,* vol. 56, pp. 1008-1021.[http://dx.doi.org/10.1016/j.rser.2015.12.009]

[84] K. Ko, S. Han, and K. S. Dan, "Performance-based settlement of frequency regulation for electric vehicle aggregators", *IEEE Trans. Smart Grid,* vol. PP, no. 99, pp. 1-1.[http://dx.doi.org/10.1109/TSG.2016.2570242]

[85] S. Gao, and H. Jia, "Integrated Configuration and Optimization of Electric Vehicle Aggregators for Charging Facilities in Power Networks With Renewables", *IEEE Access,* vol. 7, pp.

84690-84700.[http://dx.doi.org/10.1109/ACCESS.2019.2924029]

[86] *De verbruiksprofielen van huishoudelijke en niethuishoudelijke elektriciteitsverbruikers voor het jaar,* .http://www.vreg.be

[87] H. Tu, H. Fengma, S. Srdic, and S. Lukic, "Extreme Fast Charging of Electric Vehicles: A Technology Overview", *IEEE Trans. electrification,* vol. 5, no. 4, pp. 861-878.[http://dx.doi.org/10.1109/TTE.2019.2958709]

[88] S. Babaei, C. Zhao, and L. Fan, "A Data-Driven Model of Virtual Power Plants in Day-Ahead Unit Commitment", *IEEE Trans. Power Syst.,* vol. 34, no. 6, p. •••.[http://dx.doi.org/10.1109/TPWRS.2018.2890714]

[89] E. Mashhour, and S.M. Moghaddas-Tafreshi, "Bidding strategy of virtual power plant for participating in energy and spinning reserve markets – part I: problem formulation", *IEEE Trans. Power Syst.,* vol. 26, no. 2, pp. 949-956.[http://dx.doi.org/10.1109/TPWRS.2010.2070884]

[90] J. Qiu, J. Zhao, H. Yang, and Z.Y. Dong, "Optimal scheduling for prosumers in coupled transactive power and gas systems", *IEEE Trans. Power Syst.,* vol. 33, no. 2, pp. 1970-1980.[http://dx.doi.org/10.1109/TPWRS.2017.2715983]

[91] L. Jian, Y. Zheng, X. Xiao, and C.C. Chan, "Optimal scheduling for vehicle-to-grid operation with stochastic connection of plug-in electric vehicles to smart grid", *Appl. Energy,* vol. 146, pp. 150-161.[http://dx.doi.org/10.1016/j.apenergy.2015.02.030]

[92] C.N. Kristien, and E. Haesen, "The Impact of Charging Plug-In Hybrid Electric Vehicles on a Residential Distribution Grid", *IEEE Trans. Power Syst.,* vol. 25, pp. 371-380.[http://dx.doi.org/10.1109/TPWRS.2009.2036481]

[93] L. Yvkoff, *Will DC Fast Charging Harm Electric Car Batteries?,* .https://www.cnet.com/roadshow/news/will-dc-fast-charging-harm-electric-car-batteries/

[94] http://www.fleetcarma.com/dc-fast-charging-guide/

[95] https://en.wikipedia.org/wiki/Tesla_Model_3

[96] *SAE Electric Vehicle and Plug-in Hybrid Electric Vehicle Conductive Charge Coupler,* SAE Standard J1772, .

[97] "Electric Vehicle Infrastructure Installation Guide, Pacific Gas and Electric Company",

[98] M. Yilmaz, and T. Philip Krein, "Review of charging power levels and infrastructure for plug-in electric and hybrid vehicles", *IEEE Int. Electr. Vehicle Conf. (IEVC),* pp. 1-8.Greenville, SC, USA[http://dx.doi.org/10.1109/IEVC.2012.6183208]

[99] A. Ahmad, M.S. Alam, and R. Chabaan, "A comprehensive review of wireless charging technologies for electric vehicles", *IEEE Trans. Transportation Electrification,* vol. 4, no. 1, pp. 38-63.[http://dx.doi.org/10.1109/TTE.2017.2771619]

[100] *What is CHAdeMO,* .http://chademo.com/01_What_is_CHAdeMO.html

[101] "Wireless Power Transfer for Light-Duty Plug-in/Electric Vehicles and Alignment Methodology",

[102] S. Li, and C.C. Mi, "Wireless power transfer for electric vehicle applications", *IEEE J. Emerg. Sel. Top. Power Electron.,* vol. 3, no. 1, pp. 4-17.[http://dx.doi.org/10.1109/JESTPE.2014.2319453]

[103] A. Kurs, A. Karalis, R. Moffatt, J.D. Joannopoulos, P. Fisher, and M. Soljacic, "Wireless power transfer via strongly coupled magnetic resonances", *Science,* vol. 317, no. 5834, pp. 83-86.[http://dx.doi.org/10.1126/science.1143254] [PMID: 17556549]

[104] A.P. Sample, D.A. Meyer, and J.R. Smith, "Analysis, experimental results, and range adaptation of magnetically coupled resonators for wireless power transfer", *IEEE Trans. Ind. Electron.,* vol. 58, no. 2, pp. 544-554.[http://dx.doi.org/10.1109/TIE.2010.2046002]

[105] X. Mou, D.T. Gladwin, R. Zhao, and H. Sun, "Survey on magnetic resonant coupling wireless power

transfer technology for electric vehicle charging", *IET Power Electron.,* vol. 12, no. 12, pp. 3005-3020.[http://dx.doi.org/10.1049/iet-pel.2019.0529]

[106] H. Kim, "Coil design and measurements of automotive magnetic resonant wireless charging system for high-efficiency and low magnetic field leakage", *IEEE Trans. Microw. Theory Tech.,* vol. 64, no. 2, pp. 383-400.[http://dx.doi.org/10.1109/TMTT.2015.2513394]

[107] Y. Nagatsuka, N. Ehara, Y. Kaneko, S. Abe, and T. Yasuda, "Compact contactless power transfer system for electric vehicles", *Proc. IPEC,* pp. 807-813.[http://dx.doi.org/10.1109/IPEC.2010.5543313]

[108] V. Cirimele, M. Diana, F. Freschi, and M. Mitolo, "Inductive power transfer for automotive applications state-of-the-art and future trends", *IEEE Trans. Industry App,* .[http://dx.doi.org/10.1109/TIA.2018.2836098]

[109] A.A.S. Mohamed, S. An, and O. Mohammed, "Coil design optimization of power pad in IPT system for electric vehicle applications", *IEEE Trans. Magn.,* vol. 54, pp. 1-5.[http://dx.doi.org/10.1109/TMAG.2017.2784381]

[110] S. Wang, K. Izaki, I. Hirota, H. Yamashita, H. Omori, and M. Nakaoka, "Induction-heated cooking appliance using new quasi-resonant ZVS-PWM inverter with power factor correction", *IEEE Trans. Ind. Appl.,* vol. 34, no. 4, pp. 705-712.[http://dx.doi.org/10.1109/28.703961]

[111] F. Sanz, C. Sagues, and S. Llorente, "Induction heating appliance with a mobile double-coil inductor", *IEEE Trans. Ind. Appl.,* vol. 51, no. 3, pp. 1945-1952.[http://dx.doi.org/10.1109/TIA.2014.2367136]

[112] C.S. Wang, O.H. Stielau, and G.A. Covic, "Design considerations for a contactless electric vehicle battery charger", *IEEE Trans. Ind. Electron.,* vol. 52, no. 5, pp. 1308-1314.[http://dx.doi.org/10.1109/TIE.2005.855672]

[113] X. Qu, H. Han, S.C. Wong, C.K. Tse, and W. Chen, "Hybrid IPT topologies with constant current or constant voltage output for battery charging applications", *IEEE Trans. Power Electron.,* vol. 30, no. 11, pp. 6329-6337.[http://dx.doi.org/10.1109/TPEL.2015.2396471]

[114] M. Budhia, J.T. Boys, G.A. Covic, and C.Y. Huang, "Development of a single-sided flux magnetic coupler for electric vehicle IPT charging systems", *IEEE Trans. Ind. Electron.,* vol. 60, no. 1, pp. 318-328.[http://dx.doi.org/10.1109/TIE.2011.2179274]

[115] M. Etezadi-Amoli, K. Choma, and J. Stefani, "Rapid-charge electric vehicle stations", *IEEE Trans. Power Deliv.,* vol. 25, no. 3, pp. 1883-1887.[http://dx.doi.org/10.1109/TPWRD.2010.2047874]

[116] Y. Ma, T. Houghton, A. Cruden, and D. Infield, "Modeling the benefits of vehicle-to-grid technology to a power system", *IEEE Trans. Power Syst.,* vol. 27, no. 2, pp. 1012-1020.[http://dx.doi.org/10.1109/TPWRS.2011.2178043]

[117] X. Wang, and Q. Liang, "Energy management strategy for plug-in hybrid electric vehicles via bidirectional vehicle-to-grid", *Systems Journal IEEE,* no. 99, pp. 1-10.[http://dx.doi.org/10.1109/11196.2007.900242]

[118] M. Kesler, M.C. Kisacikoglu, and L.M. Tolbert, "Vehicle-to-grid reactive power operation using plug-in electric vehicle bidirectional offboard charger", *IEEE Trans. Ind. Electron.,* vol. 61, pp. 6778-6784.[http://dx.doi.org/10.1109/TIE.2014.2314065]

[119] U.K. Madawala, M. Neath, and D.J. Thrimawithana, "A Power–frequency controller for bidirectional inductive power transfer systems", *IEEE Trans. Ind. Electron.,* vol. 60, no. 1, pp. 310-317.[http://dx.doi.org/10.1109/TIE.2011.2174537]

[120] D.J. Thrimawithana, U.K. Madawala, and M. Neath, "A synchronization technique for bidirectional IPT systems", *IEEE Trans. Ind. Electron.,* vol. 60, no. 1, pp. 301-309.[http://dx.doi.org/10.1109/TIE.2011.2174536]

[121] U.K. Madawala, and D.J. Thrimawithana, "A bidirectional inductive power interface for electric vehicles in V2G systems", *IEEE Trans. Ind. Electron.,* vol. 58, no. 10, pp. 4789-4796.[http://dx.doi.org/10.1109/TIE.2011.2114312]

[122] Y. Jang, and M.M. Jovanovic, "A Contactless Electrical Energy Transmission System for Portable-Telephone Battery Chargers", *IEEE Trans. Ind. Electron.,* vol. 50, no. 3, pp. 520-527.[http://dx.doi.org/10.1109/TIE.2003.812472]

[123] G.B. Joun, and B.H. Cho, "An energy transmission system for an artificial heart using leakage inductance compensation of transcutaneous transformer", *IEEE Trans. Power Electron.,* vol. 13, no. 6, pp. 1013-1022.[http://dx.doi.org/10.1109/63.728328]

[124] Q. Xu, D. Hu, B. Duan, and J. He, "A fully implantable stimulator with wireless power and data transmission for experimental investigation of epidural spinal cord stimulation", *IEEE Trans. Neural Syst. Rehabil. Eng.,* vol. 23, no. 4, pp. 683-692.[http://dx.doi.org/10.1109/TNSRE.2015.2396574] [PMID: 25680207]

[125] S. Jeong, Y.J. Jang, and D. Kum, "Economic Analysis of the Dynamic Charging Electric Vehicle", *IEEE Trans. Power Electron.,* vol. 30, no. 11, pp. 6368-6377.[http://dx.doi.org/10.1109/TPEL.2015.2424712]

[126] G. Nagendra, L. Chen, G.A. Covic, and J.T. Boys, "Detection of EVs on IPT highways", *IEEE J. Emerg. Sel. Top. Power Electron.,* vol. 2, no. 3, pp. 584-597.[http://dx.doi.org/10.1109/JESTPE.2014.2308307]

[127] P. Athalye, D. Maksimovic, and R. Erickson, "High-performance front-end converter for avionics applications", *IEEE Trans. Aerosp. Electron. Syst.,* vol. 39, no. 2, pp. 462-470.[http://dx.doi.org/10.1109/TAES.2003.1207258]

[128] T. Bieler, M. Perrottet, V. Nguyen, and Y. Perriard, "Wireless power and data transmission", *IEEE Trans. Ind. Electron.,* vol. 38, no. 5, pp. 1266-1272.

[129] J. Wu, C. Zhao, Z. Lin, J. Du, Y. Hu, and X. He, "Wireless power and data transfer via a common inductive link using frequency division multiplexing", *IEEE Trans. Ind. Electron.,* vol. 62, no. 12, pp. 7810-7820.[http://dx.doi.org/10.1109/TIE.2015.2453934]

[130] Y. Sun, P.X. Yan, Z.H. Wang, and Y.Y. Luan, "The parallel transmission of power and data with the shared channel for an inductive power transfer system", *IEEE Trans. Power Electron.,* vol. 31, no. 8, pp. 5495-5502.[http://dx.doi.org/10.1109/TPEL.2015.2497739]

[131] S.J. Huang, T.S. Lee, and T.H. Huang, "Inductive power transfer systems for PT-based ozone-driven circuit with flexible capacity operation and frequency-tracking mechanism", *IEEE Trans. Ind. Electron.,* vol. 61, no. 12, pp. 6691-6699.[http://dx.doi.org/10.1109/TIE.2014.2316247]

[132] Z.H. Wang, Y.P. Li, Y. Sun, C.S. Tang, and X. Lv, "Load detection model of voltage-fed inductive power transfer system", *IEEE Trans. Power Electron.,* vol. 28, no. 11, pp. 5233-5243.[http://dx.doi.org/10.1109/TPEL.2013.2243756]

[133] A.M. Bradley, M.D. Feezor, H. Singh, and F.Y. Sorrell, "Power systems for autonomous underwater vehicles", *IEEE J. Oceanic Eng.,* vol. 26, no. 4, pp. 526-538.[http://dx.doi.org/10.1109/48.972089]

[134] H. Singh, "Docking for an autonomous ocean sampling network", *IEEE J. Oceanic Eng.,* vol. 26, no. 4, pp. 498-514.[http://dx.doi.org/10.1109/48.972084]

[135] K.W. Klontz, D.M. Divan, D.W. Novotny, and R.D. Lorenz, "Contactless power delivery system for mining applications", *IEEE Trans. Ind. Appl.,* vol. 31, no. 1, pp. 27-35.[http://dx.doi.org/10.1109/28.363053]

[136] G.A.J. Elliott, G.A. Covic, D. Kacprzak, and J.T. Boys, "A new concept: Asymmetrical pick-ups for inductively coupled power transfer monorail systems", *IEEE Trans. Magn.,* vol. 42, no. 10, pp. 3389-3391.[http://dx.doi.org/10.1109/TMAG.2006.879619]

[137] A. Kawamura, K. Ishioka, and J. Hirai, "Wireless transmission of power and information through one high-frequency resonant AC link inverter for robot manipulator applications", *IEEE Trans. Ind. Appl.,* vol. 32, no. 3, pp. 503-508.[http://dx.doi.org/10.1109/28.502160]

[138] *"hapter 3: EV Charging Technology: Conductive and Wireless," Wireless Charging Technology and*

the Future of Electric Transportation., SAE, .

[139] https://www.raeng.org.uk/publications/other/23-wind-turbine

[140] http://documents.worldbank.org/curated/en/2015/03/24365411/solarresource-mapping-pakistan-solar-modeling-report

[141] https://news.energysage.com/what-are-the-most-efficient-solar-panels-on-the-market/

[142] https://www.epa.gov/greenpower/benefits-using-green-power

[143] https://ourworldindata.org/grapher/energy-consumption-by-source-and-region?country=~OWID_WRL

[144] https://www.autocarpro.in/news-international/covid19hit-global-auto-inc-to-see-deeper-dec-ine-than-during-the-great-recession-lmc-55993

[145] Y.K. Wu, C.C. Huang, and C.L. Lin, "resolution of the unit commitment problems by using the hybrid taguchi-ant colony system algorithm", *Int. J. Electr. Power Energy Syst.*, vol. 49, no. 3, pp. 188-198.

[146] Q.P. Zheng, J. Wang, and A.L. Liu, "Stochastic optimization for unit commitment-a review", *IEEE Trans. Power Syst.*, vol. 30, no. 4, pp. 1913-1924.[http://dx.doi.org/10.1109/TPWRS.2014.2355204]

[147] Y. Zhang, J. Wang, and B. Zeng, "Chance-constrained two-stage unit commitment under uncertain load and wind power output using bilinear benders decomposition", *IEEE Trans. Power Syst.*, vol. 32, no. 5, pp. 3637-3647.[http://dx.doi.org/10.1109/TPWRS.2017.2655078]

[148] A.M. Nanda, and S. Suri, "Some new findings on automatic generation control of an interconnected hydrothermal system with conventional controllers", *IEEE Trans. Energ. Convers.*, vol. 21, no. 1, pp. 187-194.[http://dx.doi.org/10.1109/TEC.2005.853757]

[149] X. Chen, "A unified frequency-domain model for automatic generation control assessment under wind power uncertainty", *IEEE Trans. Smart Grid,* vol. 1, no. 1, pp. 159-173.

[150] R. Patel, C. Li, L. Meegahapola, B. McGrath, and X. Yu, "Enhancing Optimal Automatic Generation Control in a Multi-Area Power System With Diverse Energy Resources", *IEEE Trans. Power Syst.*, vol. 34, no. 5, pp. 3465-3475.[http://dx.doi.org/10.1109/TPWRS.2019.2907614]

[151] M.A. Ortega-Vazquez, and D.S. Kirschen, "Estimating the spinning reserve requirements in systems with significant wind power generation penetration", *IEEE Trans. Power Syst.*, vol. 24, no. 1, pp. 114-124.[http://dx.doi.org/10.1109/TPWRS.2008.2004745]

[152] S.S. Reddy, P.R. Bijwe, and A.R. Abhyankar, "Joint energy and spinning reserve market clearing incorporating wind power and load forecast uncertainties", *IEEE Syst. J.*, vol. 9, no. 1, pp. 152-164.[http://dx.doi.org/10.1109/JSYST.2013.2272236]

[153] T. Rodrigues, P.J. Ramírez, and G. Strbac, "Risk-averse bidding of energy and spinning reserve by wind farms with on-site energy storage", *IET Renew. Power Gener.*, vol. 12, no. 2, pp. 165-173.[http://dx.doi.org/10.1049/iet-rpg.2017.0223]

[154] https://evadoption.com/ev-statistics-of-the-week-range-price-and-battery-size-of-current-y-available-in-the-us-bevs/

[155] H. Louie, "K. Strunz, "Hierarchical Multiobjective Optimization for Independent System Operators (ISOs) in Electricity Markets", *IEEE Trans. Power Syst.*, vol. 21, no. 4, pp. 1583-1591.[http://dx.doi.org/10.1109/TPWRS.2006.881151]

[156] R. Singh, P.R. Kumar, and L. Xie, "The ISO problem: Decentralized stochastic control via bidding schemes", *Proc. IEEE 53rd Annu. Allerton Conf. Commun. Control Comput.*, pp. 552-559.[http://dx.doi.org/10.1109/ALLERTON.2015.7447053]

[157] M. Zang, M. Clemens, C. Cimala, J. Streckert, and B. Schmuelling, "Simulation of inductive power transfer systems exposing a human body with two-step scaled-frequency FDTD methods", *IEEE Trans. Magn.*, vol. 53, no. 6, p. •••.[http://dx.doi.org/10.1109/TMAG.2017.2654798]

[158] B.G. Choi, Y. Sohn, E.S. Lee, S.H. Han, H.R. Kim, and C.T. Rim, "Coreless Transmitting Coils with Conductive Magnetic Shield for Wide-Range Ubiquitous IPT", *IEEE Trans. Power Electron.,* vol. 34, no. 3, pp. 2539-2552.[http://dx.doi.org/10.1109/TPEL.2018.2839107]

[159] A.A.S. Mohamed, A. Meintz, P. Schrafel, and A. Calabro, "Testing and Assessment of EMFs and Touch Currents From 25-kW IPT System for Medium-Duty EVs", *IEEE Trans. Vehicular Technol.,* vol. 68, pp. 7477-7487.[http://dx.doi.org/10.1109/TVT.2019.2920827]

[160] A. Babu, and B. George, "Sensor system to aid the vehicle alignment for inductive EV chargers", *IEEE Trans. Ind. Electron.,* vol. 66, no. 9, pp. 7338-7346.[http://dx.doi.org/10.1109/TIE.2018.2883193]

[161] V-B. Vu, V-T. Phan, M. Dahidah, and V. Pickert, "Multiple output inductive charger for electric vehicles", *IEEE Trans. Power Electron.,* vol. 34, no. 8, pp. 7350-7368.[http://dx.doi.org/10.1109/TPEL.2018.2882945]

[162] D.C. Corrêa, U.C. Resende, and F.S. Bicalho, "Experiments with a compact wireless power transfer system using strongly coupled magnetic resonance and metamaterials", *IEEE Trans. Magn.,* vol. 55, no. 8, p. •••.[http://dx.doi.org/10.1109/TMAG.2019.2913767]

[163] S. Huang, Z. Li, Y. Li, X. Yuan, and S. Cheng, "A Comparative Study Between Novel and Conventional Four-Resonator Coil Structures in Wireless Power Transfer", *IEEE Trans. Magn.,* vol. 50, no. 11, pp. 1-4.[http://dx.doi.org/10.1109/TMAG.2014.2331962]

[164] W. Zhou, S. Sandeep, P. Wu, P. Yang, and S.Y. Huang, "A wideband strongly coupled magnetic resonance wireless power transfer system and its circuit analysis", *IEEE Microw. Wirel. Compon. Lett.,* vol. 28, no. 12, pp. 1152-1154.[http://dx.doi.org/10.1109/LMWC.2018.2876767]

[165] J. Pries, V.P.N. Galigekere, O.C. Onar, and G. Su, "A 50-kw three-phase wireless power transfer system using bipolar windings and series resonant networks for rotating magnetic fields", *IEEE Trans. Power Electron.,* vol. 35, no. 5, pp. 4500-4517.[http://dx.doi.org/10.1109/TPEL.2019.2942065]

[166] Y. Jiang, L. Wang, Y. Wang, J. Liu, X. Li, and G. Ning, "Analysis Design and Implementation of Accurate ZVS Angle Control for EV's Battery Charging in Wireless High Power Transfer", *IEEE Trans. Ind. Electron.,* vol. 66, no. 5, pp. 4075-4085.[http://dx.doi.org/10.1109/TIE.2018.2795523]

[167] T. Campi, S. Cruciani, F. Maradei, and M. Feliziani, "Near Field reduction in a Wireless Power Transfer System using LCC compensation", *IEEE Trans. Electromagn. Compat.,* vol. 59, no. 2, pp. 686-694.[http://dx.doi.org/10.1109/TEMC.2016.2641383]

[168] Q.H. Liu, "A pseudospectral frequency-domain (PSFD) method for computational electromagnetics", *IEEE Antennas Wirel. Propag. Lett.,* vol. 1, pp. 131-134.[http://dx.doi.org/10.1109/LAWP.2002.806755]

[169] R. Kastner, "Half-Way Duality in Electromagnetics Using an Explicit Expression for the Half-Curl Operator", *IEEE trans. on antennas and progagation,* vol. 68, no. 5, pp. 3747-3750.[http://dx.doi.org/10.1109/TAP.2020.2963938]

[170] B.E. MacNeal, J.R. Brauer, and R.N. Coppolino, "A general finite element vector potential formulation of electromagnetics using a time-integrated electric scalar potential", *IEEE Trans. Magn.,* vol. 26, no. 5, pp. 1768-1770.[http://dx.doi.org/10.1109/20.104518]

[171] H.M. Schey, "div, grad, curl, and all that, an informal text on vector calculus", 3rd ed

[172] http://mswebtutor.com/gradient-directional-derivative-divergence-and-curl-of-function-in-hindi/

[173] http://www.mysearch.org.uk/website1/html/197.VecCalc.html

[174] K.H. Wolfgang, *Panofsky and Melba Phillips, CLASSICAL ELECTRICITY MAGNETISM.,* 2nd edAddison-Wesley publishing Co., Inc., .2nd ed

[175] K. David, "Cheng, Addison, Field and Wave Electromagnetics", *Wesley.* ISBN 978-0-201-52820-6

[176] T. Fawwaz, "Ulaby, Umberto Ravaioli, Pearson, Fundamentals of Applied Electromagnetics", 7th

Global Edition

[177] W. Hayt, *John Buck, Engineering Electromagnetics.,* 7th edMc Graw Hill, .7th ed

[178] G.A. Covic, and J.T. Boys, "Inductive power transfer", *Proc. IEEE,* vol. 101, no. 6, pp. 1276-1289.[http://dx.doi.org/10.1109/JPROC.2013.2244536]

[179] S.Y. Choi, B.W. Gu, and S.Y. Jeong, "Advances in wireless power transfer systems for roadway-powered electric vehicles", *IEEE J. Emerg. Sel. Top. Power Electron.,* vol. 3, no. 1, pp. 18-36.[http://dx.doi.org/10.1109/JESTPE.2014.2343674]

[180] W. Zhang, S.C. Wong, and C.K. Tse, "Design for efficiency optimization and voltage controllability of series-series compensated inductive power transfer systems", *IEEE Trans. Power Electron.,* vol. 29, no. 1, pp. 191-200.[http://dx.doi.org/10.1109/TPEL.2013.2249112]

[181] M. Pinuela, D.C. Yates, and S. Lucyszyn, "Maximizing DC-to-load efficiency for inductive power transfer", *IEEE Trans. Power Electron.,* vol. 28, pp. 2437-2447.[http://dx.doi.org/10.1109/TPEL.2012.2215887]

[182] B. Esteban, M. Sid-Ahmed, and N.C. Kar, "A comparative study of power supply architectures in wireless EV charging systems", *IEEE Trans. Power Electron.,* vol. 30, no. 11, pp. 6408-6422.[http://dx.doi.org/10.1109/TPEL.2015.2440256]

[183] H. Feng, and S. M. Lukic, "Reduced-Order Modeling and Design of Single-Stage LCL Compensated IPT System for Low Voltage Vehicle Charging Applications", *IEEE Trans on Vehicular Technology,* vol. 69, no. 4, pp. 3728-3739.[http://dx.doi.org/10.1109/TVT.2020.2972201]

[184] F. Lu, H. Zhang, H. Hofmann, W. Su, and C. Mi, "A dual-coupled LCC-compensated IPT system with a compact magnetic coupler", *IEEE Trans. Power Electron.,* vol. 33, no. 7, pp. 6391-6402.[http://dx.doi.org/10.1109/TPEL.2017.2748391]

[185] H. Feng, S. Duan, and J. Zhao, "An LCC-compensated resonant converter optimized for robust reaction to large coupling variation in dynamic wireless power transfer", *IEEE Trans. Ind. Electron.,* vol. 63, no. 10, pp. 6591-6601.[http://dx.doi.org/10.1109/TIE.2016.2589922]

[186] M. Kim, H.P. Park, and J.H. Jung, *Practical Design Methodology of IH and IPT Dual-Functional Apparatus.,* .[http://dx.doi.org/10.1109/TPEL.2020.2976054]

[187] A. Karalis, *Novel photonic phenomena in nanostructured material systems with applications and Mid-range efficient insensitive Wireless Energy-Transfer,* Massachusetts Institute of Technology, . All rights reserved

[188] A. J. Pointo, and H. M. Howarth, "AC and DC Network Theory", SPRINGERSCIENCE+BUSINES S MEDIA, B.V

[189] J. DAVID IRWIN, and R. MARK NELMS, "Basic Engineering Circuit Analysis, 11th", *WILEY.*

[190] T.D.P. Perera, D.N.K. Jayakody, S.K. Sharma, S. Chatzinotas, and J. Li, *Simultaneous wireless information and power transfer (SWIPT): Recent advances and future challenges, .*

[191] J. Garnica, R.A. Chinga, and J. Lin, "Wireless power transmission: from far field to near field", *Proc. IEEE,* vol. 101, no. 6, pp. 1321-1331.[http://dx.doi.org/10.1109/JPROC.2013.2251411]

[192] J. Kim, J. Kim, and S. Kong, "Coil Design and Shielding Methods for a Magnetic Resonant Wireless Power Transfer System", *Proc. IEEE,* vol. 101, no. 6, pp. 1332-1342. [J].[http://dx.doi.org/10.1109/JPROC.2013.2247551]

[193] M. Lu, and K.D.T. Ngo, "Analytical calculation of proximity-effect resistance for planar coil with litz wire and ferrite plate in inductive power transfer", *IEEE Trans. Ind. Appl.,* vol. •••, p. 1.[http://dx.doi.org/10.1109/TIA.2018.2890366]

[194] D. Patil, M. Mcdonough, J. Miller, B. Fahimi, and P.T. Balsara, "Wireless power transfer for vehicular applications: Overview and challenges", *IEEE Trans. Transport. Electrific,* vol. 4, no. 1, pp. 3-37.[http://dx.doi.org/10.1109/TTE.2017.2780627]

[195] T. Kan, T.D. Nguyen, J.C. White, R.K. Malhan, and C. Mi, "A new integration method for an electric vehicle wireless charging system using LCC compensation Topology", *IEEE Trans. Power Electron.,* vol. 32, no. 2, pp. 1638-1650.[http://dx.doi.org/10.1109/TPEL.2016.2552060]

[196] J.H. Kim, "Development of 1-MW inductive power transfer system for a high-speed train", *IEEE Trans. Ind. Electron.,* vol. 62, no. 10, pp. 6242-6250.[http://dx.doi.org/10.1109/TIE.2015.2417122]

[197] S.K. Nobar, J.M. Niya, and B.M. Tazehkand, "Performance analysis of cognitive wireless powered communication networks under unsaturated traffic condition", *IEEE Trans. Green Commun. Netw.,* vol. 4, no. 3, pp. 819-831.[http://dx.doi.org/10.1109/TGCN.2020.2978264]

[198] A. Zakerian, S. Vaez-Zadeh, and A. Babaki, "A dynamic WPT system with high efficiency and high power factor for electric vehicles", *IEEE Trans. Power Electron.,* vol. 35, no. 7, pp. 6732-6740.[http://dx.doi.org/10.1109/TPEL.2019.2957294]

[199] Y. Chen, B. Yang, O. Li, H. Feng, X. Zhou, Z. He, and R. Mai, "Reconfigurable Topology for IPT System Maintaining Stable Transmission Power Over Large Coupling Variation", *IEEE Trans. Power Electron.,* vol. 35, no. 5, pp. 4915-4924.[http://dx.doi.org/10.1109/TPEL.2019.2946778]

[200] S. Samanta, and A. K. Rathore, "Analysis and Design of Load Independent ZPA Operation for P/S, PS/S, P/SP and PS/SP Tank Networks in IPT Applications", *IEEE Trans. Power Electron.* Early access article[http://dx.doi.org/10.1109/TPEL.2018.2794623]

[201] Y. Yao, Y. Wang, X. Liu, F. Lin, and D. G. Xu, "A Novel Parameter Tuning Method for Double-sided LCL Compensated WPT System with Better Comprehensive Performance", *IEEE Trans. Power Electron,* vol. PP, no. 99, p. 1-1.[http://dx.doi.org/10.1109/TPEL.2017.2778255]

[202] Y. Zhang, Z. Yan, Z. Liang, S. Li, and C.C. Mi, "A high-power wireless charging system using LCL-N topology to achieve a compact and lowcost receiver", *IEEE Trans. Power Electron.,* vol. 35, no. 1, pp. 131-137.[http://dx.doi.org/10.1109/TPEL.2019.2914363]

[203] Y. Chen, H. Zhang, C.S. Shin, C.H. Jo, S.J. Park, and D.H. Kim, "An Efficiency Optimization-Based Asymmetric Tuning Method of Double-Sided LCC Compensated WPT System for Electric Vehicles", *IEEE Trans. Power Electron.,* vol. 35, no. 11, pp. 11475-11487.[http://dx.doi.org/10.1109/TPEL.2020.2984712]

[204] W. Han, K.T. Chau, C. Jiang, W. Liu, and W.H. Lam, "Design and analysis of quasi-omnidirectional dynamic wireless power transfer for fly-and-charge", *IEEE Trans. Magn.,* vol. 55, no. 7, pp. 1-9.[http://dx.doi.org/10.1109/TMAG.2019.2895716]

[205] S. Zhou, and C. Chris, "Multi-Paralleled LCC Reactive Power Compensation Networks and Their Tuning Method for Electric Vehicle Dynamic Wireless Charging", *IEEE Trans. Ind. Electron.,* vol. 63, no. 10, p. •••.[http://dx.doi.org/10.1109/TIE.2015.2512236]

[206] Z. Zhang, H. Pang, A. Georgiadis, and C. Cecati, "Wireless power transfer—An overview", *IEEE Trans. Ind. Electron.,* vol. 66, no. 2, pp. 1044-1058.[http://dx.doi.org/10.1109/TIE.2018.2835378]

[207] J. Lu, G. Zhu, D. Lin, Y. Zhang, J. Jiang, and C.C. Mi, "Unified Load-Independent ZPA Analysis and Design in CC and CV Modes of Higher Order Resonant Circuits for WPT Systems", *IEEE Trans. Transport. Electrific.,* vol. 5, no. 4, pp. 977-987.[http://dx.doi.org/10.1109/TTE.2019.2940337]

[208] Y. Jiang, "Analysis, design, and implementation of accurate ZVS angle control for EV battery charging in wireless highpower transfer", *IEEE Trans. Ind. Electron.,* vol. 66, no. 5, pp. 4075-4085.[http://dx.doi.org/10.1109/TIE.2018.2795523]

[209] H. Li, S. Chen, J. Fang, Y. Tang, and M. de Rooij, "A low-subharmonic, full-range, and rapid pulse density modulation strategy for ZVS full-bridge converters", *IEEE Trans. Power Electron.,* vol. 34, no. 9, pp. 8871-8881.[http://dx.doi.org/10.1109/TPEL.2018.2889116]

[210] Y. Zhang, Z. Yan, T. Kan, X. Zeng, S. Chen, and C.C. Mi, "Modeling and analysis of a strongly coupled series–parallel-compensated wireless power transfer system", *IEEE J. Emerg. Sel. Top. Power Electron.,* vol. 7, no. 2, pp. 1364-1370.[http://dx.doi.org/10.1109/JESTPE.2018.2860935]

[211] Z. Pantic, S. Bai, and S.M. Lukic, "ZCS LCC-compensated resonant inverter for inductive-powe--transfer application", *IEEE Trans. Ind. Electron.,* vol. 58, no. 8, pp. 3500-3510.[http://dx.doi.org/10.1109/TIE.2010.2081954]

[212] Y. Chen, R. Mai, Y. Zhang, M. Li, and Z. He, "Improving misalignment tolerance for IPT system using a third-coil", *IEEE Trans. Power Electron.*.[http://dx.doi.org/10.1109/TPEL.2018.2867919]

[213] R. Mai, B. Yang, Y. Chen, N. Yang, Z. He, and S. Gao, "A misalignment tolerant IPT system with intermediate coils for constant-current output", *IEEE Trans. Power Electron.,* vol. 34, no. 8, pp. 7151-7155.[http://dx.doi.org/10.1109/TPEL.2019.2898686]

[214] S. Kim, G. Covic, and J.T. Boys, "Comparison of tripolar and circular pads for IPT charging systems", *IEEE Trans. Power Electron.,* vol. 33, no. 7, pp. 6093-6103.[http://dx.doi.org/10.1109/TPEL.2017.2740944]

[215] M. Budhia, G. Covic, and J. Boys, "Design and optimization of circular magnetic structures for lumped inductive power transfer systems", *IEEE Trans. Power Electron.,* vol. 26, no. 11, pp. 3096-3108.[http://dx.doi.org/10.1109/TPEL.2011.2143730]

[216] T.C. Beh, M. Kato, and T. Imura, "Automated impedance matching system for robust wireless power transfer via magnetic resonance coupling", *IEEE Trans. Ind. Electron.,* vol. 60, no. 9, pp. 3689-3698.[http://dx.doi.org/10.1109/TIE.2012.2206337]

[217] T. Imura, and Y. Hori, "Maximizing air gap and efficiency of magnetic resonant coupling for wireless power transfer using equivalent circuit and Neumann formula", *IEEE Trans. Ind. Electron.,* vol. 58, no. 10, pp. 4746-4752.[http://dx.doi.org/10.1109/TIE.2011.2112317]

[218] A.P. Sample, D.T. Meyer, and J.R. Smith, "Analysis Experimental Results and Range Adaptation of Magnetically Coupled Resonators for Wireless Power Transfer", *IEEE Trans. Ind. Electron.,* vol. 58, no. 2, pp. 544-554.[http://dx.doi.org/10.1109/TIE.2010.2046002]

[219] R. Huang, B. Zhang, D. Qiu, and Y. Zhang, "Frequency Splitting Phenomena of Magnetic Resonant Coupling Wireless Power Transfer", *IEEE Trans. Magn.,* vol. 50, no. 11, pp. 1-4.[http://dx.doi.org/10.1109/TMAG.2014.2331143]

[220] A.P. Sample, B.H. Waters, S.T. Wisdom, and J.R. Smith, "Enabling seamless wireless power delivery in dynamic environments", *Proc. IEEE,* vol. 101, no. 6, pp. 1343-1358.[http://dx.doi.org/10.1109/JPROC.2013.2252453]

[221] *Standard SAEJ2954,* .http://standards.sae.org/j2954_201605/

[222] J.M. Carrasco, L.G. Franquelo, and J.T. Bialasiewicz, "Power electronic systems for the grid integration of renewable energy sources: a survey", *IEEE Trans. Ind. Electron.,* vol. 53, no. 4, pp. 1002-1016.[http://dx.doi.org/10.1109/TIE.2006.878356]

[223] Md. Abdus Salam, *Fundamentals of Electrical Power Systems Analysis,* .[http://dx.doi.org/10.1007/978-981-15-3212-2]

[224] F. Blaabjerg, Z. Chen, and S.B. Kjaer, "Power electronics as efficient interface in dispersed power generation system", *IEEE Trans. Power Electron.,* vol. 19, no. 5, pp. 1184-1194.[http://dx.doi.org/10.1109/TPEL.2004.833453]

[225] B. Singh, and R. Kushwaha, "A PFC based EV battery charger using a bridgeless isolated SEPIC converter", *IEEE Trans. Ind. Appl.,* vol. 56, no. 1, pp. 477-487.[http://dx.doi.org/10.1109/TIA.2019.2951510]

[226] M.M. Jovanovic, and Y. Jang, "State-of-the-art single-phase active power-factor-correction techniques for high-power applications - an overview", *IEEE Trans. Ind. Electron.,* vol. 52, no. 3, pp. 701-708.[http://dx.doi.org/10.1109/TIE.2005.843964]

[227] C.E. Kim, D.k. Yang, J.B. Lee, and G.W. Moon, "High-efficiency two-inductor PFC boost converter employing SPDT relay", *IEEE Trans. Power Electron.,* vol. 30, no. 6, pp. 2901-2904.[http://dx.doi.org/10.1109/TPEL.2014.2368567]

[228] C. Chan, and K. Chau, "An overview of power electronics in electric vehicles", *IEEE Trans. Ind. Electron.,* vol. 44, no. 1, pp. 3-13.[http://dx.doi.org/10.1109/41.557493]

[229] Y. Sun, Y. Liu, M. Su, W. Xiong, and J. Yang, "Review of active power decoupling topologies in single-phase systems", *IEEE Trans. Power Electron.,* vol. 31, no. 7, pp. 4778-4794.

[230] H.J. Kim, H. Hirayama, S. Kim, K.J. Han, R. Zhang, and J.W. Choi, *Review of Near-field Wireless Power and Communication for Biomedical Applications.,* vol. Vol. 5, IEEE Access, .

[231] C.C. Huang, C.L. Lin, and Y-K. Wu, "Simultaneous wireless power/data transfer for electric vehicle charging", *IEEE Trans. Ind. Electron.,* vol. 64, no. 1, pp. 682-690.[http://dx.doi.org/10.1109/TIE.2016.2608765]

[232] C.C. Huang, and C.L. Lin, "Wireless power and bidirectional data transfer scheme for battery charger", *IEEE Trans. Power Electron.,* vol. 33, no. 6, pp. 4679-4689.[http://dx.doi.org/10.1109/TPEL.2017.2725940]

[233] S-M. Kim, J-I. Moon, I-K. Cho, J-H. Yoon, W-J. Byun, and H-C. Choi, "Advanced Power Control Scheme in Wireless Power Transmission for Human Protection From EM Field", *IEEE Trans. Microw. Theory Tech.,* vol. 63, no. 3, p. •••.[http://dx.doi.org/10.1109/TMTT.2015.2398444]

[234] S.A.E. Taskforce, http://www.greencarcongress.com/2012/01/j2954-20120122.html

[235] C-C. Huang, C-L. Lin, J-J. Kao, J-J. Chang, and G-J. Sheu, "Vehicle parking guidance for wireless charge using GMR sensors", *IEEE Trans. Vehicular Technol.,* vol. 67, no. 8, pp. 6882-6894.[http://dx.doi.org/10.1109/TVT.2018.2827069]

[236] J.M. Miller, P.T. Jones, and J.M. Li, "ORNL experience and challenges facing dynamic wireless power charging of EV's", *IEEE Circuits Syst. Mag.,* vol. 15, no. 2, pp. 40-53.[http://dx.doi.org/10.1109/MCAS.2015.2419012]

[237] T. Fujita, T. Yasuda, and H. Akagi, "A dynamic wireless power transfer system applicable to a stationary system", *IEEE Trans. Ind. Appl.,* vol. 53, no. 4, pp. 3748-3757.[http://dx.doi.org/10.1109/TIA.2017.2680400]

[238] S. Choi, J. Huh, W.Y. Lee, S.W. Lee, and C.T. Rim, "New cross-segmented power supply rails for roadway-powered electric vehicles", *IEEE Trans. Power Electron.,* vol. 28, no. 12, pp. 5832-5841.[http://dx.doi.org/10.1109/TPEL.2013.2247634]

[239] https://solargis.com/maps-and-gis-data/download/pakistan

[240] A.G. Bhuiyan, K. Sugita, A. Hashimoto, and A. Yamamoto, "InGaN solar cells: present state of the art and important challenges", *IEEE J. Photovolt.,* vol. 2, pp. 276-293.[http://dx.doi.org/10.1109/JPHOTOV.2012.2193384]

[241] M. Yamaguchi, and A. Luque, "High efficiency and high concentration in photovoltaics", *IEEE Trans. Electron Dev.,* vol. 46, no. 10, pp. 2139-2144.[http://dx.doi.org/10.1109/16.792009]

[242] https://energyeducation.ca/encyclopedia/Solar_cell_efficiency

[243] V. Yaramasu, B. Wu, and P.C. Sen, "High-power wind energy conversion systems: state-of-the-art and emerging technologies", *Proc. IEEE,* vol. 103, no. 5, pp. 740-788.[http://dx.doi.org/10.1109/JPROC.2014.2378692]

[244] L. Xie, P.M.S. Carvalho, L.A.F.M. Ferreira, B.H. Krogh, N. Popli, and M.D. Ilić, "Wind integration in power systems: operational challenges and possible solutions", *Proc. IEEE,* vol. 99, pp. 214-232.[http://dx.doi.org/10.1109/JPROC.2010.2070051]

[245] Xu1 Wanghuai, Zheng Huanxi, Liu. Yuan, Zhou Xiaofeng, Zhang Chao, Song Yuxin, Deng Xu, Leung Michael, Yang Zhengbao, X. Xu Ronald, Lin Wang Zhong, Cheng Zeng Xiao, and Wang Zuankai, "A droplet-based electricity generator with high instantaneous power density", *nature 578,* pp. 392-396.

[246] J. Nie, Z. Wang, Z. Ren, S. Li, X. Chen, and Z. Lin Wang, "Power generation from the interaction of a

liquid droplet and a liquid membrane", *Nat. Commun.,* vol. 10, no. 1, p. 2264.[http://dx.doi.org/10.1038/s41467-019-10232-x] [PMID: 31118419]

[247] X. Nian, F. Peng, and H. Zhang, "Regenerative Braking System of Electric Vehicle Driven by Brushless DC Motor", *IEEE Trans. Ind. Electron.,* vol. 61, no. 10, pp. 5798-5808.[http://dx.doi.org/10.1109/TIE.2014.2300059]

[248] K.M. Choo, and C.Y. Won, "Design and Analysis of Electrical Braking Torque Limit Trajectory for Regenerative Braking in Electric Vehicles With PMSM Drive Systems", *IEEE Trans. Power Electron.,* vol. 35, no. 12, pp. 13308-13321.[http://dx.doi.org/10.1109/TPEL.2020.2994615]

[249] I. Sheikhian, N. Kaminski, S. Vob, W. Scholz, and E. Herweg, "Optimisation of the reverse conducting IGBT for zero-voltage switching applications such as induction cookers", *IET Circuits Dev. Syst.,* vol. 8, no. 3, pp. 176-181.[http://dx.doi.org/10.1049/iet-cds.2013.0215]

[250] C. Cui, and Y. Liu, "A Full Power Range ZVS Control Technology for Bidirectional Inductive Power Transfer System", *Industrial Electronics Society (IECON) 2020 The 46th Annual Conference of the IEEE,* pp. 3861-3865.

[251] Y. Lim, "An adaptive impedance-matching network based on a novel capacitor matrix for wireless power transfer", *IEEE Trans. Power Electron,* vol. 29, no. 8, pp. 4403-4413.[http://dx.doi.org/10.1109/TPEL.2013.2292596]

[252] J. Kim, D. Kim, and Y. Park, "Analysis of Capacitive Impedance Matching Networks for Simultaneous Wireless Power Transfer to Multiple Devices", *IEEE Trans. Ind. Electron.,* vol. 62, pp. 2807-2813.[http://dx.doi.org/10.1109/TIE.2014.2365751]

[253] D. Wang, X. Qu, Y. Yao, and P. Yang, "Hybrid Inductive-Power-Transfer Battery Chargers for Electric Vehicle Onboard Charging With Configurable Charging Profile", *IEEE Trans. Intell. Transp. Syst.,* vol. 22, no. 1, pp. 592-599.[http://dx.doi.org/10.1109/TITS.2020.2976647]

[254] C.S. Wang, G.A. Covic, and O.H. Stielau, "Power transfer capability and bifurcation phenomena of loosely coupled inductive power transfer systems", *IEEE Trans. Ind. Electron.,* vol. 51, no. 1, pp. 148-157.[http://dx.doi.org/10.1109/TIE.2003.822038]

[255] J-J. Kao, C-L. Lin, Y-C. Liu, C-C. Huang, and H-S. Jian, *"Adaptive Bidirectionally Inductive Power and Data Transmission System", IEEE Trans. Power Electron Early Access.,* .

[256] D. W. Hart, "Resonant converters, Power electronics",

[257] https://sites.google.com/site/bccl606/

SUBJECT INDEX

A

B

C

D

E

Chih-Cheng Huang & Chun-Liang Lin

N

O

P

R

S

T

U

V

W

Z

Made in United States
Orlando, FL
05 November 2026

90200365R10095